MILLER
BROS
U.S.A
S.C-U.S.A

THE COMPLETE BOOK OF U.S. MILITARY POCKET KNIVES

From 1800 to the Present

Michael W. Silvey

Schiffer Publishing Ltd
4880 Lower Valley Road • Atglen, PA 19310

Other books by Michael W. Silvey include:
United States Military Knives Collectors Guide, Silvey, Boyd, 1989.
United States Military Knives 1941 to 1991, Silvey, 1992.
Knives of the United States Military in Vietnam, Silvey, 1996.
Knives of the United States Military World War II, Silvey, 1999.
U.S. Military Knives Bayonets and Machetes Price Guide, Knife World Publications, 2000.
Pocket Knives of the United States Military, Silvey, 2002.
SOG Knife Collector, Silvey, 2014.

Library of Congress Control Number: 2015930809

Book Design by Michael W. Silvey.
Cover Design by John Cheek
Type set in Times New Roman

ISBN: 978-0-7643-4827-3
Printed in China

Published by Schiffer Publishing, Ltd.
4880 Lower Valley Road
Atglen, PA 19310
Phone: (610) 593-1777; Fax: (610) 593-2002
E-mail: Info@schifferbooks.com

For our complete selection of fine books on this and related subjects, please visit our website at www.schifferbooks.com. You may also write for a free catalog.

This book may be purchased from the publisher. Please try your bookstore first.

We are always looking for people to write books on new and related subjects. If you have an idea for a book, please contact us at proposals@schifferbooks.com.

Schiffer Publishing's titles are available at special discounts for bulk purchases for sales promotions or premiums. Special editions, including personalized covers, corporate imprints, and excerpts can be created in large quantities for special needs. For more information, contact the publisher.

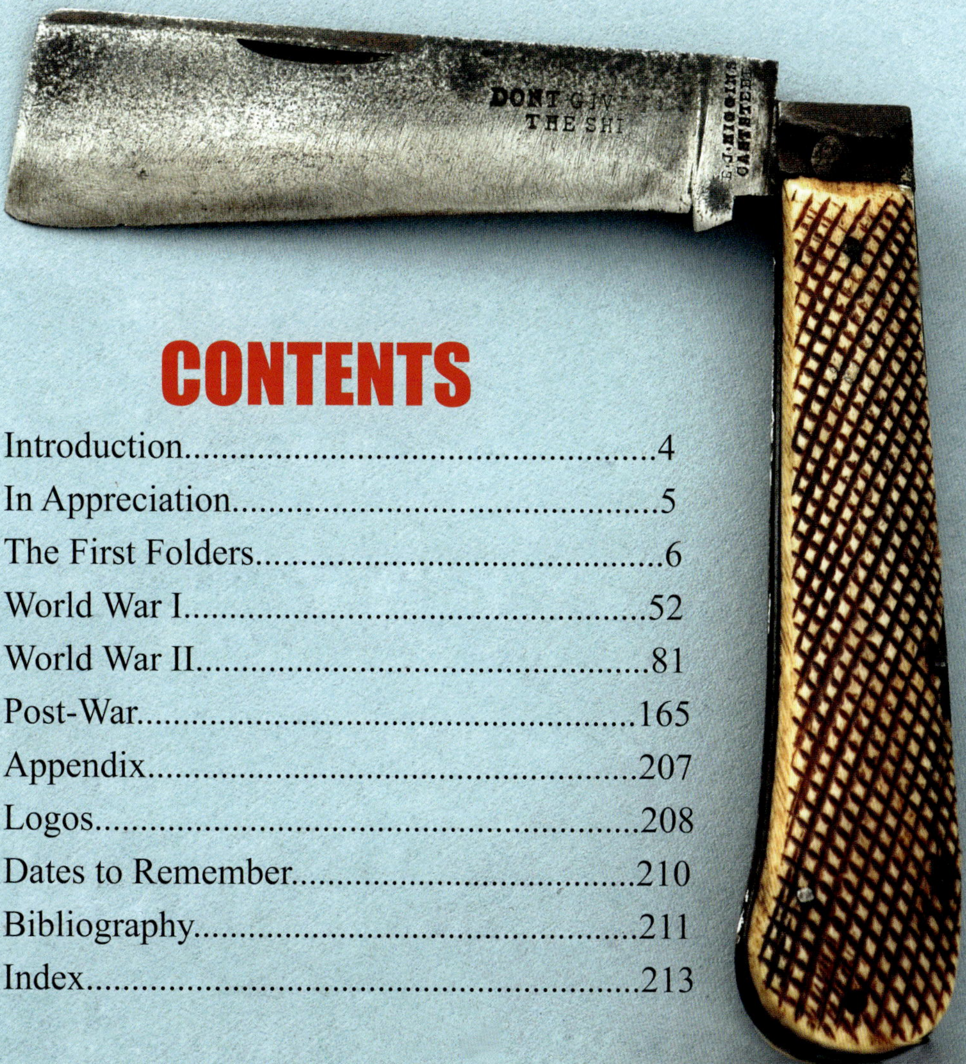

CONTENTS

INTRODUCTION

This book is a celebration of the pocket knives of the United States military. Believing in the old adage that "a picture is worth a thousand words," I have tried to provide photographs that accurately portray the knife as well as put it into context, however, dating the knives is difficult because of the lack of records and information on the older knives. Nevertheless, there are some facts that help. For example, after 1890 all U.S. cutlery imports required the country of origin stamped on the product. Cast steel in knife blades began to show up around 1875. Shear steel went out of usage around 1900. Knives with celluloid handles did not exist before 1870. The integral liner and bolster did not exist much before the Civil War. So, there are a number of factors that help to date the knives but it is far from precise, particularly so for older knives.

D.E. Henry, in the Introduction to his book on *Collins Machetes and Bowies,* points out that collecting is a relatively recent phenomena possible because we can now afford the luxury of owning things purchased for their own sake ... not for their use. What was previously the purview of museums is now open to thousands. The knives men brought home as mementos of their service or were available as government surplus are now highly prized and have taken on a value far in excess of their original cost. As more and more information becomes available the field broadens and interest increases. What does the knife tell us about the society in which it was used? What kind of a tool was it? It may have performed tasks that are no longer even thought about. What does it tell us about those that made it and those that used it? Finding the answers to these questions can be a satisfying challenge but there are other avenues in collecting. There is the interacting and sharing interests among collectors that often provides a great deal of social satisfaction. Then there is the hunt for the knives themselves. Like any hunt, this can be full of excitement. What is around the next turn? The hunt goes where you go, a stop at a flea market while traveling, a walk through an antique shop, and, of course knife and gun shows. So, although Henry was correct in that it is our standard of living that makes collecting possible, I have never met two people who collect for the same reasons. There are many books about knife collecting and if you are a beginning collector I encourage you to find out more about the hobby by reading and exploring the books.

This is a picture book that presents a small segment of knife collecting in as pleasant a light as possible. It is a book for collectors and I believe there is something in here for the beginner, as well as, advanced collector. By no means does this book exhaust the subject of U.S. military pocket knives as there are areas that are hardly explored. For example, the four blade all metal general purpose utility knives provides opportunity for exploration and study. This book also contains some knives that have never been documented before. These are primarily the navy rope knives. Despite the problems with dating the knives I have tried to organize the book chronologically and then by type of knife. I hope the book increases interest in the hobby and provides the reader/viewer with hours of enjoyment.

It should be noted that the some of the vital statistics of the knives have frequently been displayed in a small block on the page. This was done simply to avoid boring repetition. One of the terms used is OAL which stands for Over All Length. This measurement is taken with the knife closed and excludes the bail.

IN APPRECIATION

First, I offer a heartfelt thank you to Frank Trzaska, Ernie Modlin, and J. Philip Bailey, who did a pre-print review and corrected some of my false assumptions and added a number of interesting facts. Also, I owe a debt of thanks to those who have provided much information through their research, authorship, and collections, especially M. H. Cole, Bernard Levine, Frank Trzaska, Adrian Van Dyk, Bob Lowry, Garry Zalesky, Mark Zalesky, and John Goins. To those who provided me their knives and other treasured items I used as background props for the photographs, I offer my thanks. So thank you Bill Mesquita, John Gibson, Gary Boyd, Mike Squyer, and Dr. John W. Brunner. Thank you C. Houston Price for your kindness and for giving me an education in publishing along with the confidence to publish the first book. Thanks to my family for their enduring support.

THE FIRST FOLDERS

The earliest pocket knives associated with the United States military are the rope knives of the ordinary sailor. The rope knife usually has a single blade with a square point or sheepsfoot design. These knives were the working tool of those who served on the wooden-hulled sailing ships making repairs to rope, sails, and other running gear. The blades were stout and could be hammered through a heavy rope with the use of a belaying pin or wooden mallet. Almost all of these knives had a hole in the butt end of the handle or were equipped with bail or clevis so that a lanyard could be attached to prevent the knife from being lost. Nevertheless, over the years many were lost either through accident or because of extreme usage and conditions, and today these knives are very rare. The earliest rope knives were probably purchased by individual sailors. It wasn't until the Civil War that there is evidence that these knives were actually purchased by the U.S. military for issue. Almost all of these rope knives dating from the Civil War up to World War I were marked with "U.S. Navy" on the blade.

For years there has been a story circulating as to the reason these knives were made with a square point. It goes that men became quarrelsome after spending long periods at sea and fights were not uncommon. A stab wound made with a pointed knife was often fatal because of damage to a vital organ. However, a square-pointed knife was capable of cutting a man but was rarely fatal and many on board ship were capable of expert stitchery facilitating emergency care. Whether this is just romantic folklore or is, in fact, based in truth is unknown. It is known, however, that there were restrictions placed on sailors and the knives they carried. In 1866 Congress prohibited merchant seamen from wearing sheath knives.

Cast steel or crucible steel is a product that was developed in the mid-1700s by Benjamin Huntsman of Sheffield, England. Its manufacture became an art whose practice made Sheffield the knife capital of the world for over one hundred years. This high carbon steel was not the formulated steel that we can buy today in quantity at most industrial supply outlets. Rather, this was a hand cooked steel made in a small batch by a man trained over his lifetime to recognize when it contained the correct mixture of iron, carbon, and heat. A man who then could turn this mixture into an ingot of steel that possessed the correct properties for a blade. These early knives are indeed works of art, each one a hand wrought masterpiece made with the best practices and materials of the day. Although every effort was made to produce a consistently high quality knife variances are frequently found. I have a number of rope knives from the same manufacturer, some with very well-worn blades that still have great snap. I have others with full blades whose springs and blade heels are completely worn out. Although the usage and maintenance of a knife can have much to do with the wear patterns, it is also clear that the tempering of steel was an art that did not have predictable scientific consistency. Despite this, it is clear that the maker took great pride in each of these knives.

Navy rope knife circa 1800

I believe that this knife dates to around 1800. The knife is 4⅝ inches long when closed and has handles of scored bone. The only marking on this knife is on the blade center with "United States/Navy" being forged deeply into the blade. This is the earliest issue folding knife the author has seen. Perhaps "issue" is not the correct term since sailors at this time were required to purchase their knives and other personal equipment from Navy stores.

Early rope knife with "United States Navy" marked on the blade.

The figure below shows that this knife did not have integral bolsters but that they were separate and added during the construction process. The figure also shows the squared tang of the knife which opens and closes with a snap and possesses a half open stop as well.

OAL: $4\frac{5}{8}$ inches
Liners: steel
Bolsters: steel

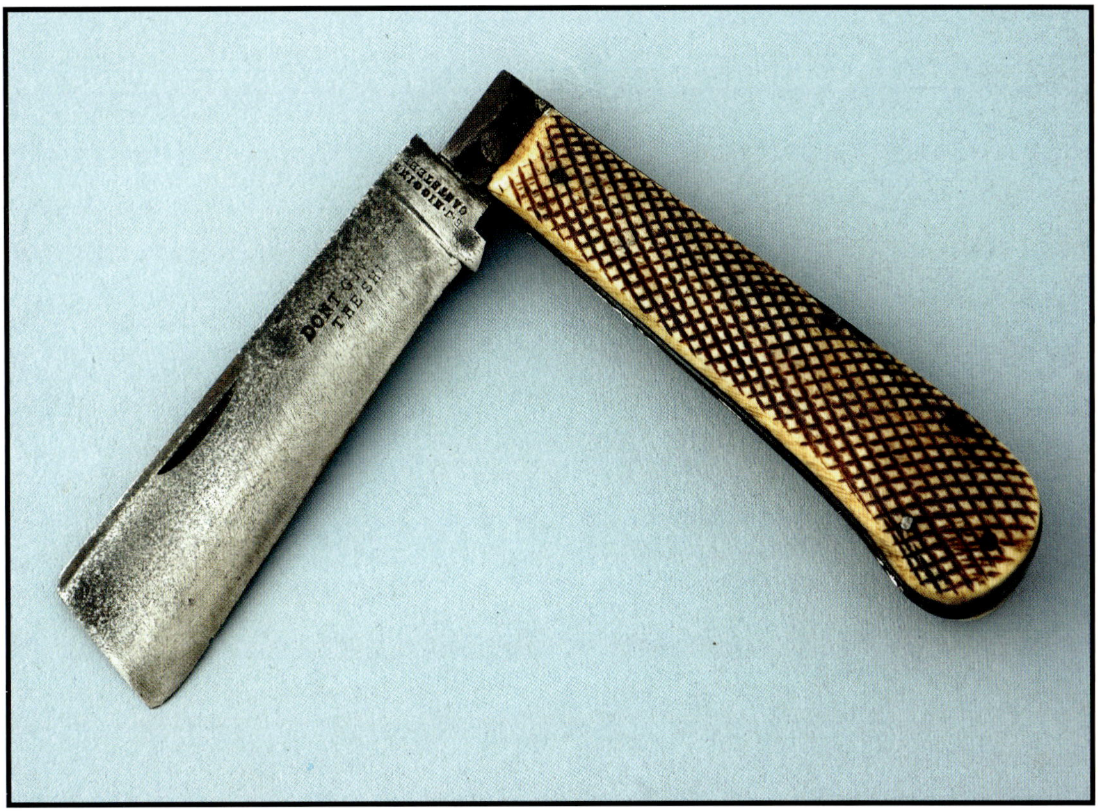

This rope knife, which probably dates from the 1820s, has handles of checkered bone and a blade marked with the maker's mark,

E.J.HIGGINS
CASTSTEEL

The blade is also marked "DON'T GIVE UP THE SHIP." These were the dying words of Captain James Lawrence aboard the USS *Chesapeake* in her fight with HMS *Shannon* on June 1, 1813, in the War of 1812. These words on a blue background became the battle flag of Commodore Oliver Hazard Perry, who in the Battle of Lake Erie on September 10, 1813, captured the British squadron, the only time in history an entire Royal Navy squadron had been captured intact. This knife is 4¼ inches closed.

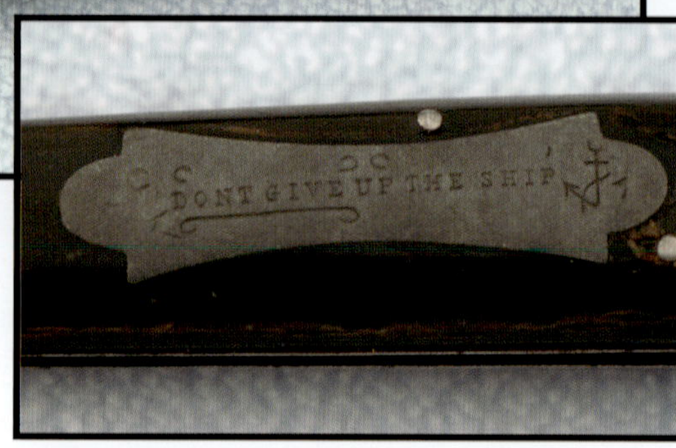

BEST ENGLISH CUTLERY

This interesting knife has horn scales and a pewter shield marked "DONT GIVE UP THE SHIP" along with decorations including an anchor. The blade ricasso is marked "BEST/ENGLISH/CUTLERY," a mark that some believe dates to the mid-19th century. This slogan, however, was generated during the War of 1812 and was popular at that time so my thinking is that the knife might be dated somewhat earlier, perhaps the 1820s. The knife measures 4 inches when closed. The liners and bolsters are iron.

This once great old rope knife was brought into this world with a square point blade but was modified by a previous owner, apparently many, many

years ago. The stag handles still reflect the handsome beauty of this old tool. A close-up image of the bolsters shows that they were made integrally with the liners, that is, they are all one piece of steel. This characteristic is found in few of the early knives.

Circa 1846

WILSON HAWKSWORTH ELLISON & CO SHEFFIELD

Originally this company started in 1825 as Wilson and Hawksworth, then in 1832 they became Wilson, Hawksworth and Moss whose name appears on the knife on the previous page. In 1846 things again changed and they became Wilson, Hawksworth, Ellison & Co. a trademark which remained until 1953. I believe that this knife was made shortly after Ellison appeared in 1846.

WARNE
CHEEVER & CO

This is a very robust rope knife and one of the very few made with hardwood handles. I believe this knife dates from around 1850 but there is slight evidence to support that opinion.

OAL: $4\frac{11}{16}$ inches
Liners: steel
Bolsters: steel

J. COX & SONS

J. Cox and Sons/Celebrated/Cutlery is stamped into the tang of this solid old rope knife. The blade center is marked U.S. "Navy" with the periods up near the center of the letters. There is no country of origin marked on the knife so it was made prior to 1890 and I believe it was made in England around the middle of the 19th century. It is 4⅝ inches long when closed. The stag is beautiful and the entire knife is quite robust.

J. Cox, circa 1850

Circa 1850

THOMAS
TURNER&Co
SHEFFIELD

This is the mark found on this Turner rope knife. The blade is also etched, "U S NAVY." Thomas Turner & Co. was established about 1802 and continued in Sheffield until the mid-1900s. Turner rope knives are noted for their dramatically curved spine.

OAL: 4¾ inches
Liners: steel
Bolsters: steel

Joseph Jeffries
circa 1850

The tang stamps of these two knives are different. The earlier knife does not include the "Sheffield" inscription. The upper knife is $4\frac{3}{4}$ inches long while the lower knife is $4\frac{9}{16}$ inches long. The frame on the upper knife is not only longer but is also more substantial.

GEORGE WOSTENHOLMS I·XL CUTLERY

This is the mark of one of the venerable cutlery firms of Sheffield, England. Wostenholm began making cutlery in the 1740s and remained in business until 1971. This knife was probably made around 1860. Wostenholm is a manufacturer noted for its high quality and this knife reflects that reputation.

OAL: 4¾ inches
Liners: steel
Bolsters: steel

Circa 1860

George Wostenholm
Circa 1860

GEORGE WOSTENHOLM I·XL CUTLERY

Both these knives are by Wostenholm and exhibit the quality attributed to that brand. The heavily used knife on the bottom still has plenty of snap. Despite their age and use both remain handsome tools.

GEORGE
WOSTENHOLM
SHEFFIELD

Retaining its well-worn original Navy issue lanyard, this Wostenholm rope knife displays a different marker's mark than the previous knives which are, I believe, a product of an earlier generation. Those Wostenholm knives shown on the previous pages display integral bolsters and do not have the Sheffield designation.

Circa 1860

TALLY- HO
C.T. BINGHAM
SHEFFIELD

OAL: $4\frac{11}{16}$ inches
Liners: steel
Bolsters: steel

According to Goins, Tally-Ho was a trademark used by F. Fenney of Sheffield, England, until about 1860 when the company was taken over by C. T. Bingham. Bingham had an agent in New York City for American sales. Some of their knives were purchased by the United States Navy. This knife has had the blade point reshaped.

Circa 1862

H.H.TAYLOR
&BROTHER
SHEFFIELD

OAL: 4½ inches
Liners: steel
Bolsters: steel

The knife above is a variation of the H. H. Taylor & Brother which is stamped with a small "U.S. NAVY" mark and a closely spaced marking and two different styles of typeface. In *Goins' Encyclopedia of Cutlery Markings*, he indicates that the H.H. Taylor only existed during 1862 and 1863 and that they maintained an office in New York City.

Circa 1862

H.H. TAYLOR & BROTHER SHEFFIELD

This variation of the Taylor knife is stamped with a large "U.S. NAVY" and a very loosely spaced mark.

OAL: 4¾ inches
Liners: steel
Bolsters: steel

CHAPIN & PHIPPS
CELEBRATED
CUTLERY

Circa 1870

Chapin and Phipps manufactured cutlery in England prior to 1900. After 1890 United States tariff laws required the country of origin to be stamped on imports. This knife has no country of origin indicated so it must have been made prior to 1890.

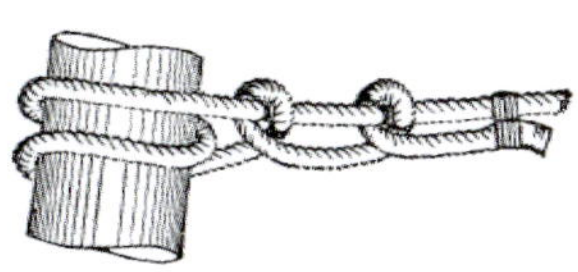
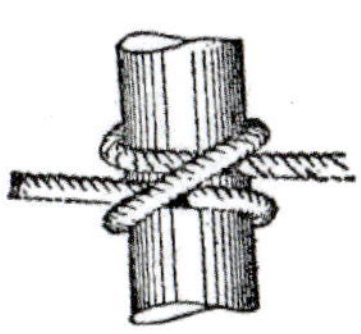
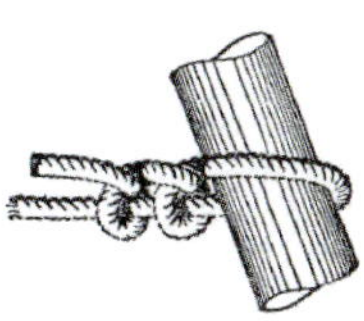

OAL: 4¾ inches
Liners: steel
Bolsters: steel

Circa 1875

F. WESTPFAL
ACME
ENGLAND

OAL: 4½ inches
Liners: steel
Bolsters: steel

F. Westpfal in New York was established in 1874 as a supplier of cutlery and related products. This knife was probably supplied to them on a contract basis from a manufacturer in England. It is the only one of these early U.S. Navy rope knives I have seen that is made with jigged bone handles.

Circa 1875

**UNION
KNIFE CO.
NAUGATUCK**

Union Knife Company in Naugatuck, Connecticut, made this large sailor's knife sometime in their history. This was one of the few early U.S. Navy rope knives not made in Sheffield, England but in the United States. It measures 4¾ inches closed and has steel liners and steel bolsters. It originally possessed a square point blade but was reshaped sometime in its earlier life. Union Knife Company was in business from 1851 to 1885.

<> R <>
RUSSELL & CO
GREEN RIVER
WORKS

Russell is one of the oldest names in American cutlery having begun operations in the 1830s. They began making pocket cutlery in 1875 and, except for a period during World War I, continued until 1941. Their rope knives are quite similar to those of other firms, however, the blades are marked NAVY KNIFE or RUSSELL/NAVY KNIFE which leads me to believe that this was not an official issue knife since it lacks the "U.S." designation. These knives are handled in wood as well as stag.

OAL: 4¼ inches
Liners: steel
Bolsters: steel

Circa 1880

It is likely that the steel bail on this U.S. Navy rope knife was added by a previous owner. Although the Navy did issue lanyards, this one also appears to have originated with a previous owner. Knowing how to work with knots and cord was a source of pride for sailors of the tall ships and this skill was often displayed in handwoven lanyards. This knife was made by Alfred Field & Company, a New York firm that dealt in hardware from the 1850s until World War II. This knife was probably a contract knife made for them by their affiliate Joseph Rodgers & Sons of Sheffield, England.

ALFRED
FIELD & Co
SHEFFIELD

OAL: 4¾ inches
Liners: steel
Bolsters: steel

Circa 1880

Circa 1885

SMITH BROTHERS

Smith Brothers located in Boston, Massachusetts, were identified in the city directory as cutlers. They were in business from 1860 to 1910. This United States Navy rope knife bearing their name must have been made for them under contract with a maker in England. The knife belonged to Joe H. Kane, pictured on the large button he had made for his mother. Note the Navy issue lanyard worn around his neck. The back of the blade tang is marked England. Kane was an enlisted man, a Landsman, on the battleship USS *Maine* which was blown up July 16, 1898 in Havana Harbor. He was one of the few who survived the explosion. The destruction of the *Maine* was the incident that ignited the Spanish-American War.

OAL: $4\frac{5}{8}$ inches
Liners: steel
Bolsters: steel

ULSTER KNIFE CO.
ELLENSVILLE, NY

The knife is a hobo pattern, which means it was designed as an entire cutlery set with knife, fork and spoon. The knife split apart to make it easy for the user. The knife is marked "Ulster Knife Co./Ellensville, NY" and is of high quality. The fork is marked "Army Knife/Ulster." The main blade is etched "Remember the Maine," which was a slogan of the Spanish American War in the late-1800s.

JOSEPH HAYWOOD & COMPY SHEFFIELD

This may be a one of a kind Navy rope knife. It appears that the main blade in this beautiful sportsman's knife has been replaced with a typically stout rope knife blade. Perhaps the main blade had been broken and the rope knife blade was a handy replacement. In any event, it makes a handsome and practical knife.

OAL: 4⅞ inches
Liners: steel
Bolsters: nickel silver

Circa 1885

H.B.CLAFLIN & COMPANY

This is the mark found on this knife which, was made around the end of the 19th century. They made knives for the United States Navy but it is not known whether they also made knives for the commercial market.

OAL: 4¾ inches
Liners: steel
Bolsters: steel

obverse side

THOMAS FENTON

reverse side

SHEFFIELD ENGLAND

is the mark found on the obverse ricasso of this knife. The reverse is stamped as shown at the left. The knife has a tapered backspring and rat tail bolsters. Little is known of this maker.

OAL: 4⅝ inches
Liners: steel
Bolsters: steel

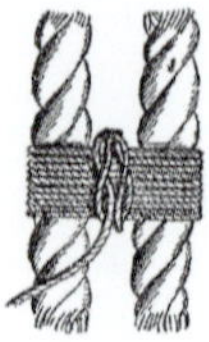

ENGLAND is the only mark found on this U.S. Navy rope knife.

Circa 1890

OAL: 4⅝ inches
Liners: steel
Bolsters: steel

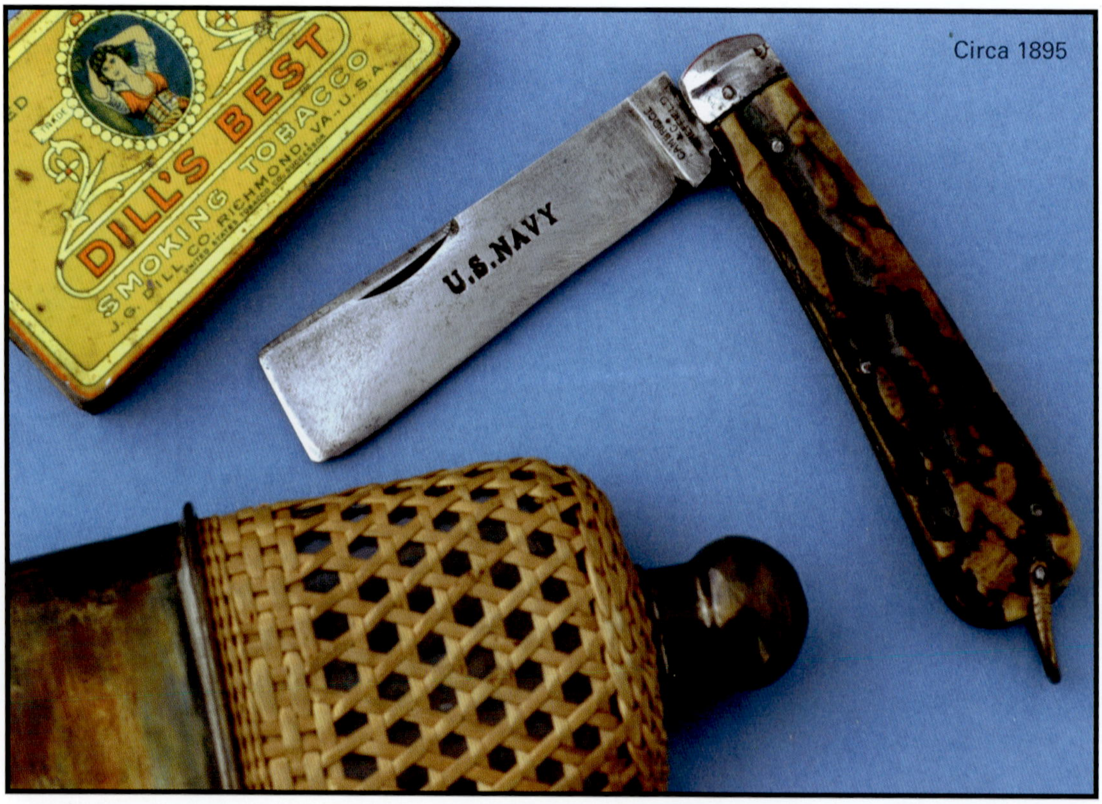

Circa 1895

CAMBRIDGE
& Co
SHEFFIELD

CAMBRIDGE
CUTLERY WORKS
SHEFFIELD

The Spanish American War took place in 1898. After the war, many of our Navy ships were painted white and this fleet was sent around the globe on a tour of goodwill and to display America's Naval might. Many of the sailors on deck probably worked with a knife like the one shown above. This one was made by Cambridge and Company of Sheffield, England, and was probably manufactured around the beginning of the twentieth century. The knife on the following page is also by Cambridge but is marked as shown at the lower left and has a brass rather than a copper bail.

OAL: 4¾ inches
Liners: steel
Bolsters: steel

U.S. NAVY

A NOTE ABOUT LANYARDS:

Rope knives were used by seamen in the everyday course of their duties such as mending sails or repairing or cutting rope. In *Knives '93* Bernard Levine had an article, "A Navy Knife of the Great War," which included a description of the *Uniform Regulations, United States Navy* that had been written in 1913. He states that each recruit, enlisted man, petty officer and chief in the seaman branch was to be issued one "Jackknife." He quotes the regulation further:

> "Knife Lanyard 232. The knife lanyard is not an article of uniform. In working dress, or when engaged in work requiring the use of a knife, the lanyard may be worn around the waist or around the neck, as most convenient."

Shown below is a U.S. Navy issue lanyard. It is made of cotton and is 40 inches in length. These were commonly used with the rope knives. In addition, many of the seamen who took pride in their ability to work with cord and knots made their own lanyards.

The knife below was made during the Civil War era by J. Ward and Company of Bronxville, New York. The knife possesses a non-tapered backspring, which is unlike almost all the other rope knives that have backsprings that gradually increase in width as they approach the butt end of the knife. The knife has rat tail bolsters and shows good quality and craftsmanship throughout.

J. WARD & Co
BRONXVILLE
NEW YORK

OAL: 4½ inches
Liners: steel
Bolsters: steel

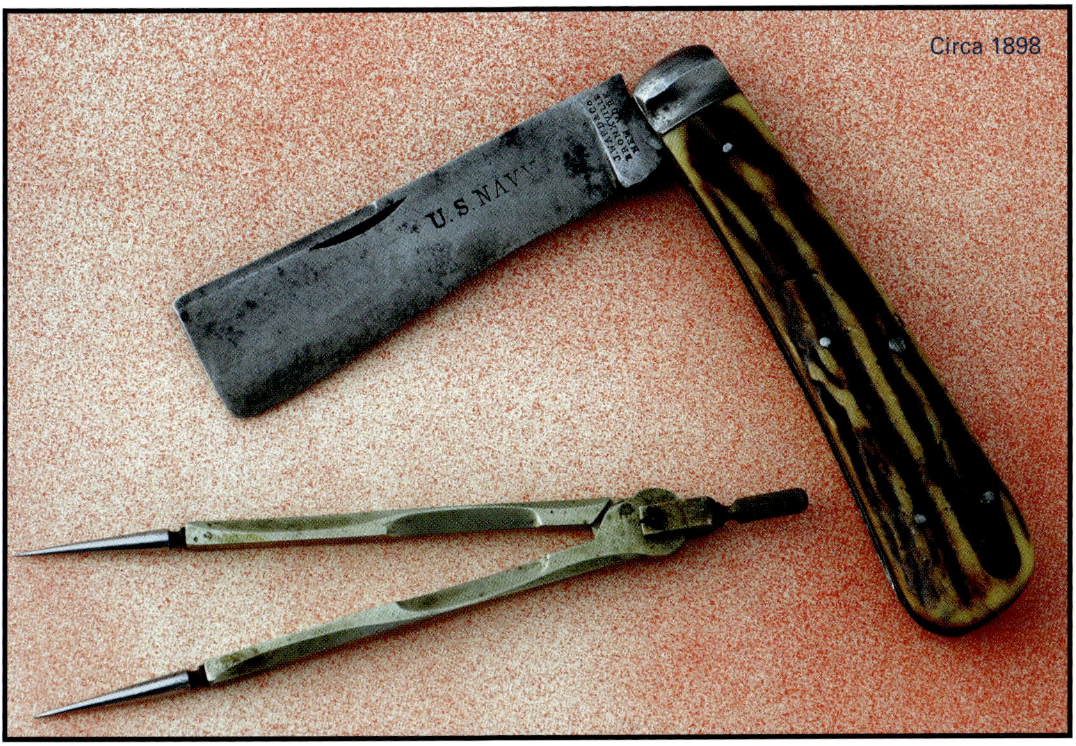

Miller Brothers started business in 1863 in Connecticut and after a couple of moves into increasingly larger space they settled in the town of Meriden in 1872. They became one of the giants of the American cutlery scene and continued making cutlery until 1926, when their focus shifted to the manufacturing of other products. One of the unique features of Miller Brothers pocketknives is that the handle scales are held in place by screws. This feature was patented by Miller Bros. I believe this gravity knife was made around the turn of the century and this model probably saw service in the Spanish American War in 1898.

Circa 1898

MILLER BROS

OAL: 4⅞ inches
Liners: brass
Bolsters: steel

Circa 1898

MILLER BROS

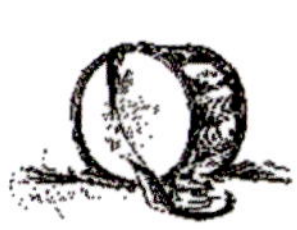

This knife is identical to the previous knife except for the bail. There are two types of lanyard attachments. I believe the bent wire brass bail on the previous knife is the earliest type and that later knives were equipped with the swivel type brass bail, which has become somewhat of a trademark of Miller Bros. rope knives. The knife shown above uses a swivel type bail.

OAL: 4⅞ inches
Liners: brass
Bolsters: steel

obverse

**ALFRED
WILLIAMS
SHEFFIELD
ENGLAND**

reverse

Alfred Williams is the mark on this U.S. Navy rope knife. This is the first of a variety of markings found on the Williams knives. The marking of the reverse is the pen nib with the "A W" on the rear most portion of the tip. Some of their knives have this pen tip on the obverse as well but most are without it altogether. This knife is also marked with the "United States Navy" in full while most just show "U.S. Navy." The lanyard to this knife is original Navy issue and measures 40 inches in length.

Circa 1890-1920

OAL: 4½ inches
Liners: steel
Bolsters: steel

Circa 1890-1920

ALFRED WILLIAMS
SHEFFIELD ENGLAND

This knife is very similar to the example on the preceding page but the pen mark is on the obverse along with the manufacturer's name and location. Alfred Williams, according to *Goins' Encyclopedia of Cutlery Markings*, was in the business of making these knives from 1890 to 1920. My experience has shown that nearly seven out of ten early U.S. Navy rope knives found today bear their mark. They had a great variety of marks.

OAL: 4½ inches
Liners: steel
Bolsters: steel

Circa 1890-1920

OAL: 4½ inches
Liners: steel
Bolsters: steel

A. Kastor & Brothers Warranted Cutlery Co. was established in the 1880s in New York City and they imported knives from Sheffield, England. Alfred Williams was among the firms that supplied knives. Kastor Brothers went on to purchase Camillus Knife Co. in 1901 which became Camillus Cutlery Company.

ALFRED WILLIAMS
SHEFFIELD, ENGLAND

A WILLIAMS
SHEFFIELD. ENG.

The mark at the left is found on the knives shown above. This is the most common marking found on rope knives by Williams. The blades of these knives are marked "U.S. NAVY," which is seen more frequently than the "UNITED STATES NAVY" marking. Another mark one may encounter, albeit seldom, is the one shown at the lower left.

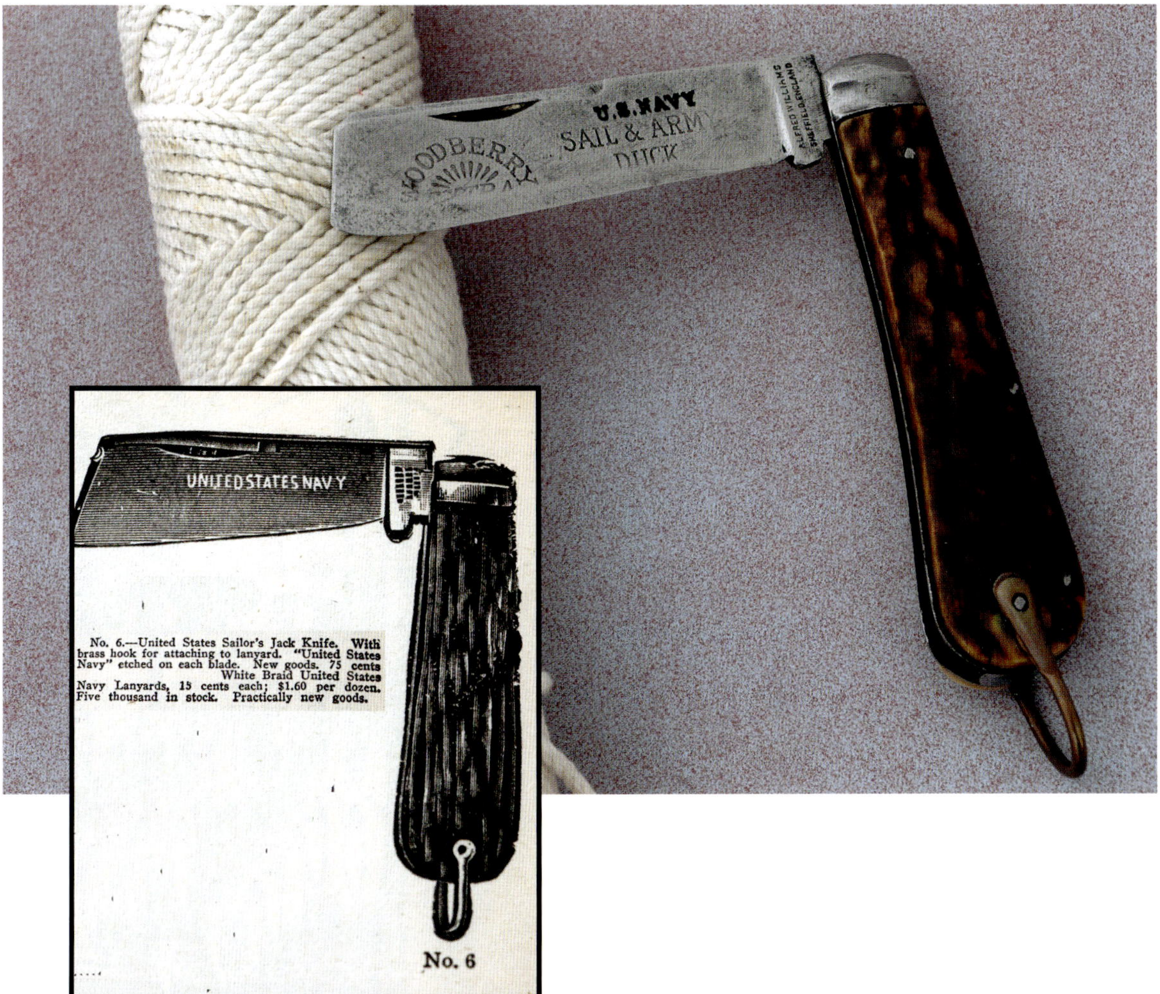

The inset was taken from the 1927 edition of the *Bannerman Catalog of Military Goods.* It serves to illustrate that many of these knives ended up as surplus property to be disposed of by the military.

The knife in the photo shows that some were given a second life as commercial incentives. This is a practice that has been observed with the Mark 1 Trench Knife of 1918 and at least one other maker's WWII rope knife. The above blade was etched, "Woodberry/Extra" and "Sail & Army/Duck."

JUNIOR ARMY & NAVY

OAL: 4½ inches
Liners: steel
Bolsters: steel

The above knife has the mark shown at the left. There is no "U. S. NAVY" mark on the blade nor is there anything else on the knife itself that would indicate that it may have been an issue knife. However, the knife is identical in manufacture and size to other marked knives. In addition, Junior Army & Navy Stores were a London-based seller of military goods and had contracts with the U.S. government. Further, they advertised in the *Stars & Stripes* during World War I as a supplier of military goods. It was also about this time that the use of the "U. S. NAVY" stamping on the blade was being discontinued. It is possible that this may have been an issue knife or it could have been privately purchased by a U.S. sailor.

Circa 1900

CHALLENGE CUTLERY Co. SHEFFIELD

OAL: 4⅝ inches
Liners: steel
Bolsters: steel

Challenge Cutlery Company of Bridgeport, Connecticut, made pocket cutlery that is highly prized among American collectors. Its roots, however, go back to 1867 when the name "Challenge" was the trademark of B.J. Eyre Company of Sheffield, England. In 1877 this firm was purchased by the Frederick Wiebusch Company of New York. Challenge Razor Works began in 1889 in Bridgeport, Connecticut, as part of the F. Wiebusch Company. In 1899 they purchased the Hatch Cutlery Company and changed the name to Challenge Cutlery Company. The knife shown above was probably made around the end of the nineteenth century.

MILLER BROS
CUT. CO
MERIDEN

A more conventional rope knife supplied by Miller Brothers Cutlery Company is shown below. Two different types of swivel lanyard loops are found on these knives. The knife on the right has the closed loop version while the knife on the left has a loop formed from brass wire. This knife was also personalized by a previous owner who did an artful job of carving his initials on the back side and incising the Navy anchor on the side shown.

OAL: 4¾ inches
Liners: steel
Bolsters: steel

MILLER BROS

This handsome knife is marked as shown at the left. These knives were made prior to World War I and few display the Navy markings. I have seen only one with the blade etched "U.S. NAVY" and another etched, "Remember The Maine." The *Maine* was an American battleship which was sunk near Havana, Cuba, in February, 1898. The sinking was caused by an unexplained explosion, however it inflamed public sentiment and "Remember the Maine" became a slogan of the soon to follow Spanish American War.

OAL: 4¾ inches
Liners: steel
Bolsters: steel

MILLER BROS	MILLER BROS CUT. CO MERIDEN	MILLER BROS CELEBRATED CUTLERY

Miller Brothers Navy rope knives are found with three different markings stamped on the ricassos. The pocket knife below is marked as shown above at the right.

OAL: 4¾ inches
Liners: steel
Bolsters: steel

PARKER & FIELD

There is little doubt that the knife below was made by Miller Brothers Cutlery. The blade, however, is marked Parker & Field, a firm that existed around the first part of the 1900s. Not much is known about Parker & Field, but I believe that they were not a manufacturer but rather sold cutlery for which they had contracted out. I am aware of another contract knife by Miller Brothers. This one has "U.S. NAVY" stamped on the blade and is marked "James R. Michael (in an arc)/MF'G CO." Unlike the Parker & Field knife below the Michael knife had the Miller patented screwed on handle scales. The James R. Michael Manufacturing Company was a military supplier.

OAL: 4¾ inches
Liners: steel
Bolsters: steel

OEHM & CO.
BALTO MD

Identical in construction to the Miller Bros. knife even to the screws holding the jigged bone scales in place, it is likely that this knife was indeed made by Miller Bros. for Oehm and Company of Baltimore, Maryland.

EASTMAN

This knife originally had a square point blade like those shown in the preceding sections. This rope knife is marked only Eastman, a brand that is unknown. The similarity to the Miller Brothers knife is obvious probably indicating that Miller Brothers was the manufacturer.

WORLD WAR I

The Great War took place between 1914 and 1918 and it is difficult to definitively state that a specific knife is pre-war or post-war. The knives that precede this section as well as those within this section may have seen service before, during, or after the war. The assignment of knives to this period is thus a bit arbitrary. Until World War I it seems that the only pocket knives attributable to the military have been the rope knives of the U.S. Navy. The use of rope knives will continue and, in fact, they are still being issued even today. Other pocket knife varieties, however, began to appear as a result of changing times. A diminutive rope knife was issued to World War I sailors as a general purpose cutting and grooming tool as it carried a manicure blade. With the advent of electricity and the radio and telephone, the electrician's knife became a practical tool to have in one's pocket. Similarly, other Signal Corps knives were issued. The Medical Department found a doctor's knife would be a handy tool to have about. America had discovered the usefulness of pocket knives. The country was becoming more settled and it was no longer necessary for the knife one carried to serve both as a weapon and a tool. Pocket knives were developed to serve stockmen, farmers, office workers, foresters, florists, vintners, factory workers, and gentlemen and the military also discovered their usefulness. This was also the time when the quality of knife making was at its peak. The fit and finish of manufactured knives was outstanding and quality was valued. During World War II there were strategic shortages of some materials and brass was not permitted in the manufacture of pocket knives. This was not the case in World War I as the knives supplied to the military during this period were made with these materials. In fact, brass was so plentiful it was exported to France as a raw material. Finally, unlike the earlier knives, almost all of these World War I knives were made here in the United States.

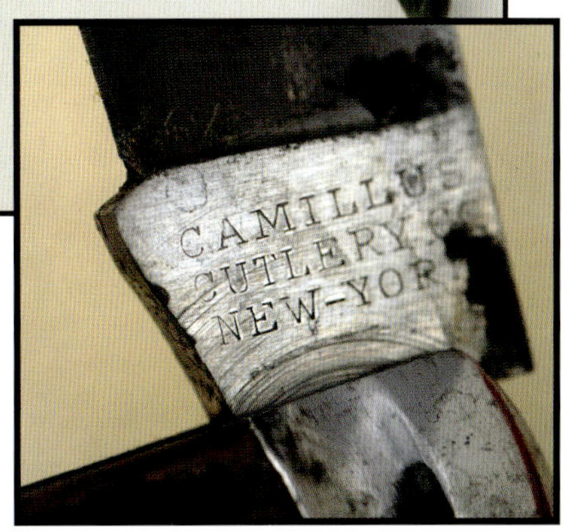

CAMILLUS
CUTLERY CO
NEW-YORK

This fine rope knife with jigged bone handles and a copper bail bears the mark shown above. This is one of the first Camillus knives supplied to the military and it was a sign of things to come. Camillus, over the years, has supplied millions of knives to the military.

OAL: $4\frac{3}{8}$ inches
Liners: steel
Bolsters: steel

Thomaston Knife Company
Thomaston, Connecticut
circa 1910

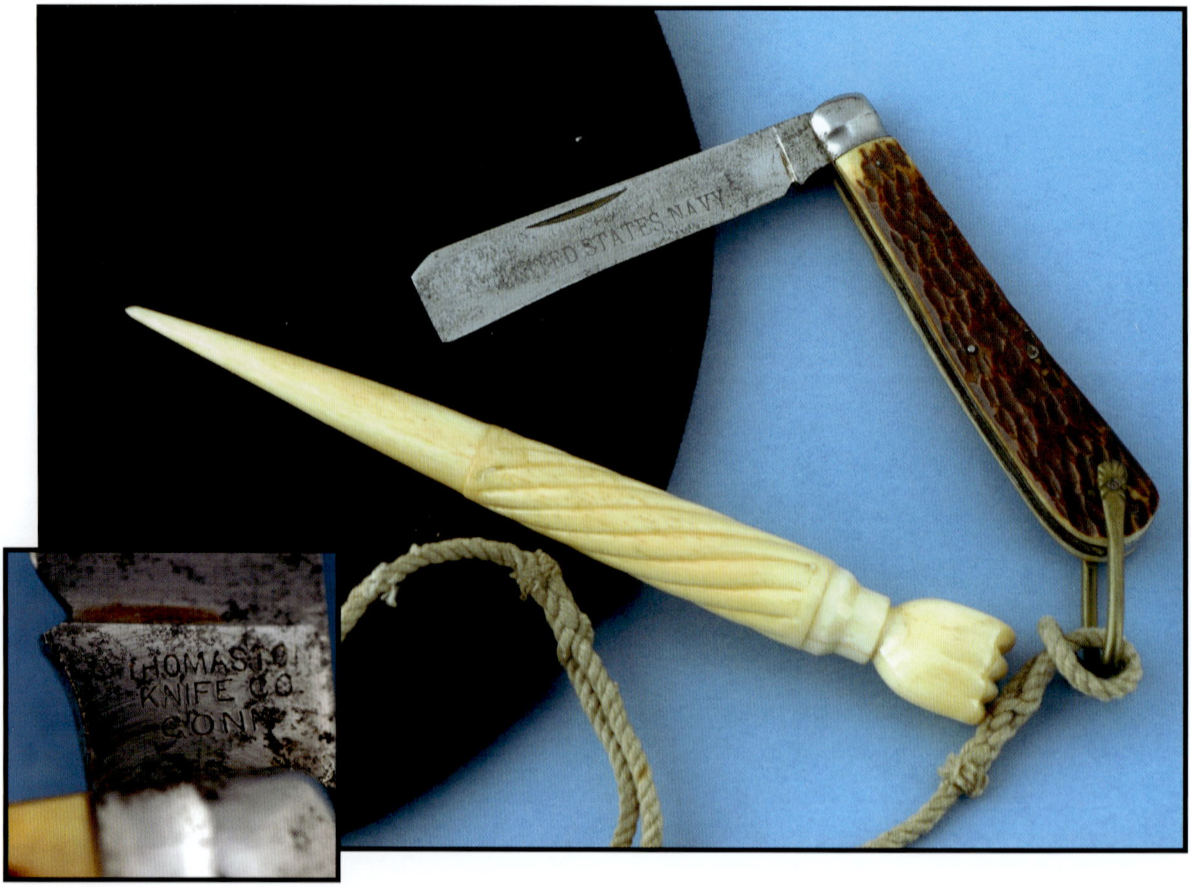

The Thomaston Knife Company of Thomaston, Connecticut, made the knife above for the Navy at the outset of World War I. This knife is nearly identical to the Camillus knife shown on the previous page except that it has a bail made of brass rather than copper.

OAL: 4⅜ inches
Liners: steel;
Bolsters: steel

The advertisement shown below was found in the Hibbard, Spencer, Bartlett & Company's 1929 Catalog. The knife shown in the catalog closely resembles the Camillus and Thomaston knives displayed on the previous pages. I have yet to find one of these HSB varieties.

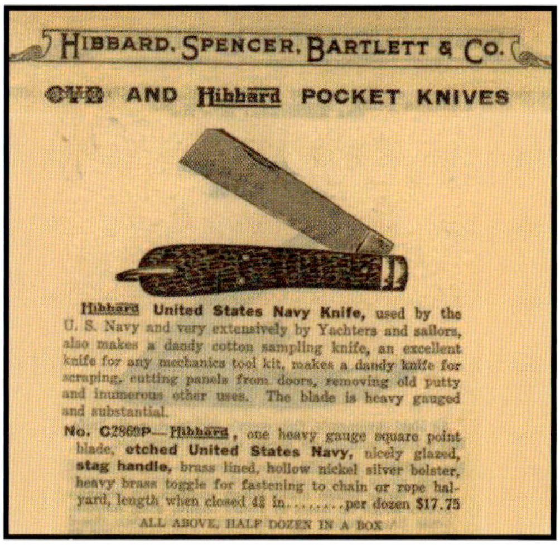

MILLER BROS CUT. CO MERIDEN

The knife with the swivel bail shown below is by Miller Brothers and differs from earlier knives of the same maker in that it is smaller. The Miller Brothers rope knife in the inset belonged to Harold Lynn Dickson, the sailor who personalized it by replacing the bone with brass scales. He did an excellent job of fitting the new brass scales and used the original screws to hold them in place. The new handle is engraved with his initials. The swivel bail has been removed.

OAL: 4⅛ inches
Liners: steel
Bolsters: steel

The knife pictured above was made by New York Knife Company of Walden, New York. The knife, of high quality, is nearly identical to the Miller Brothers rope knife. This would appear to be Navy issue but there is little evidence to support that speculation.

OAL: 4⅛ inches
Liners: brass
Bolsters: nickel silver

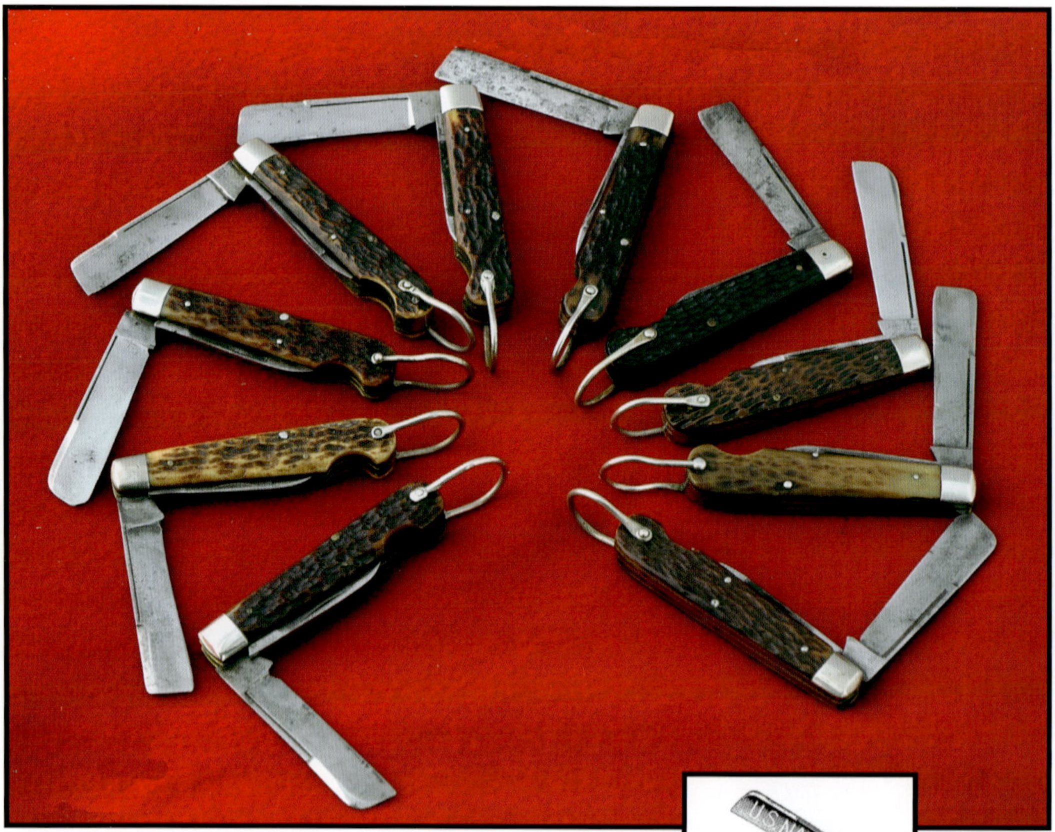

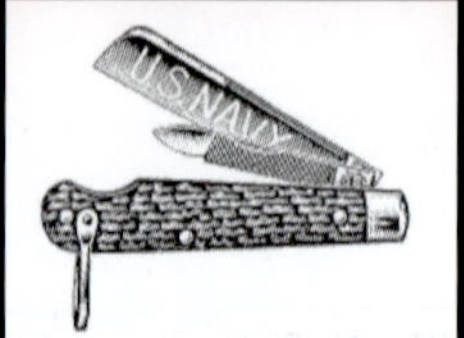

No. 40N. NEW MODEL U. S. NAVY JACK KNIFE, length open 5¾ inches. Has large blade, also small manicure blade. Loop for lanyard. Price, 60 cents.

I believe that this pattern knife came into use during World War I and was issued broadly to include personnel in all branches of service not just the Navy. The makers in this photo include clockwise from left to right: Schatt and Morgan, Robeson Cutlery Company, Valley Forge Cutlery Company, Camillus Cutlery Company, Cattaraugus Cutlery Company, Union Cutlery Company, Imperial Knife Company, New York Knife Company, Canton Cutlery Company, and finally an unmarked knife.

CAMILLUS
CUTLERY CO.
CAMILLUS, N.Y.
U.S.A.

In World War I this knife became the new knife for the Navy. It was supplied in large numbers and was produced by several manufacturers including but, perhaps, not limited to those shown on the previous page. The knives are remarkably consistent from maker to maker. They measure $3\frac{1}{4}$ inches when closed. All have jigged bone handles and a bail. They have two blades, a sheepsfoot main blade and manicure blade. The knife shown above was made by Camillus Cutlery, which supplied the Navy with about 155,000 of these.

The knife in the photo has the marking shown at the right. Schatt and Morgan were in business from 1902 to about 1928 when the company was sold to Queen Cutlery. The advertisement shown on page 58 came from the 1927 *Bannerman Catalogue of Military Goods.* The photograph of the knife in this advertisement is also shown in the Schrade Cutlery Company catalog of 1926. It shows a knife with "U.S. NAVY" etched on the blade. Many of these knives have been observed over the years but I know of none noted to have been found with such an etching.

SCHATT & MORGAN
CUTLERY CO.
TITUSVILLE PA

W.R. Case & Sons

Robeson Cutlery Co.

The top knife shown above is by W.R. Case & Sons and the knife below it was made by Robeson Cutlery Company.

for all similar Navy knives

OAL: 3¼ inches
Liners: brass
Bolsters: nickel silver

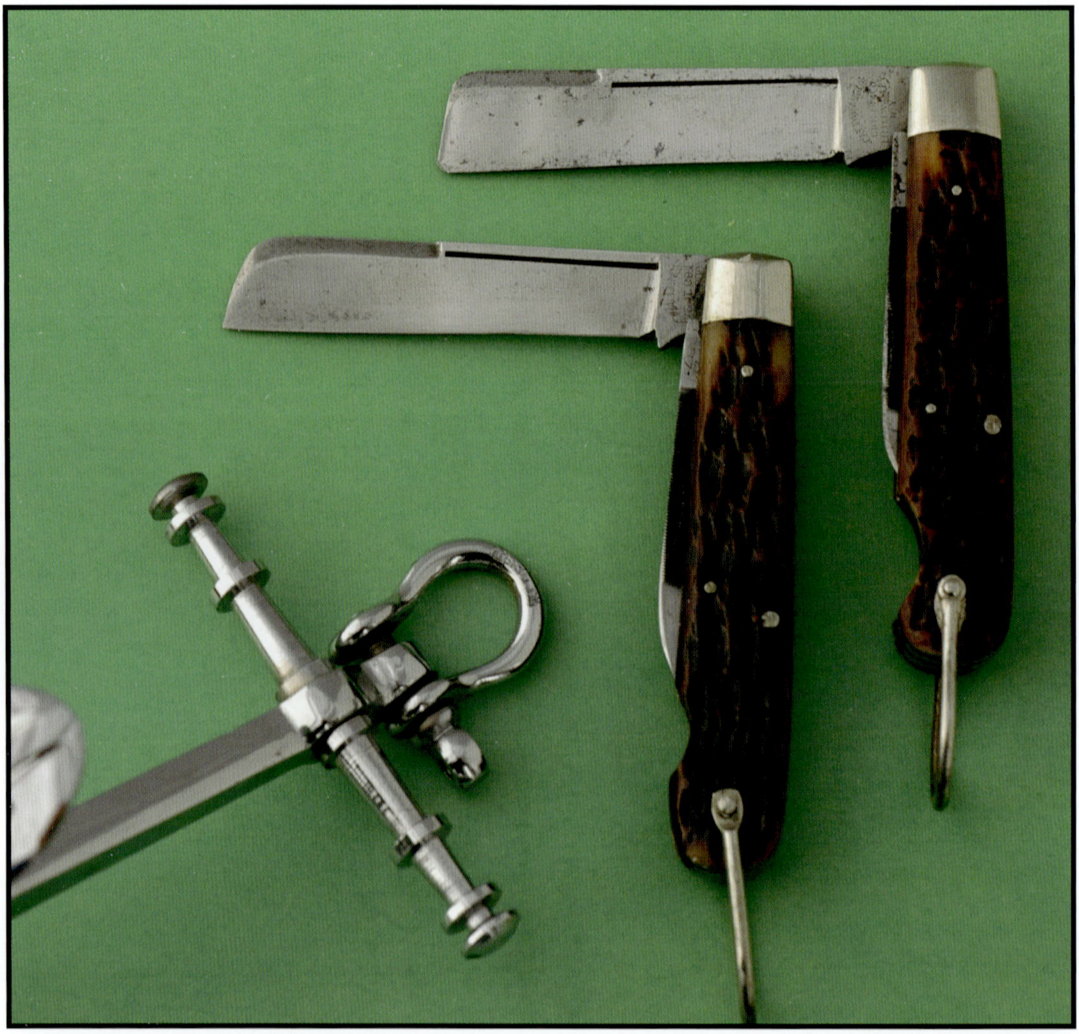

The top knife is by Cattaraugus and the knife to its left is by Challenge, both high quality manufacturers.

CATTARAUGUS
CUTLERY Co.
LITTLE VALLEY
N.Y.

This early United States Army Signal Corps easy opener pocket knife was made by Keen Kutter, E.C. Simmons Hardware Company of St. Louis, Missouri. It is a stout knife that measures $3\frac{7}{8}$ inches when closed. Handles are ebony, liners are brass and the bolsters are nickel silver.

This two-blade Signal Corp jackknife is by Miller Brothers. It is a well made, heavy duty pocket knife. The handles are wood, the bolsters are nickel silver and the liners are brass. The escutcheon plate is marked "SC-USA" for Signal Corps – United States Army. Both blades are marked with the makers name.

MILLER
BROS
U.S.A.

OAL: 3¾ inches
Liners: brass
Bolsters: nickel silver

**EMPIRE
WINSTED
CT**

Like the knife on the preceding page this easy opener was meant for use by the U.S. Army Signal Corps. This one was made by Empire/Winsted/CT.

This early Signal Corps knife is by Camillus and has their Sword Brand mark on the main blade while the small pen blade has the Camillus four line mark.

CAMILLUS
CUTLERY CO.
CAMILLUS.N.Y.
U.S.A.

CAMILLUS
CUTLERY CO.
CAMILLUS, N.Y.
U.S.A.

This three-blade cattleman's knife bears the mark shown at the left on each blade. The escutcheon plate is marked "U.S.SIGNAL CORPS." The knife, which is quite stout, measures nearly 4¼ inches closed. An identical knife can be found marked with the Keen Kutter logo.

OAL: 4¼ inches
Liners: brass
Bolsters: nickel silver

THE UNITED STATES ARMY SIGNAL CORPS POCKET INSPECTOR'S KIT

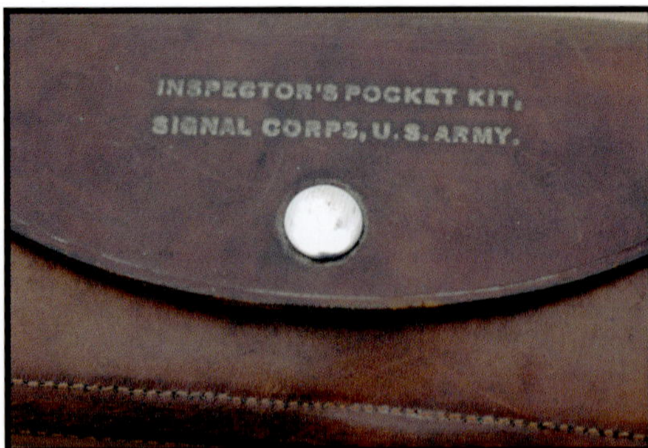

The Inspector's Pocket Kit of World War I is a bifold leather box about 6½ inches by 3½ inches by 1¾ inches. Imprinted on the flap is, "INSPECTOR'S POCKET KIT./ SIGNAL CORPS, U.S. ARMY." The box is quite well-made of rugged top grain leather and lined with suede to reinforce the boxed flaps and fold lines. Inside, there are seven loops for holding tools. All the tools are marked "S.C.U.S.A." for Signal Corps United States Army.

War Dept. Document No.541, Chapter 8, dated 1916 offers the following description from the Chief Signal Officer, Manual, #3, Technical Equipment of the Signal Corps:

INSPECTOR'S POCKET KIT.

[The inspector's pocket kit is manufactured under specification No. 186.]

As the name implies, the inspector's pocket kit is of such size that it may be conveniently placed in a pocket of clothing. The tools furnished with this

kit are only those which may be required for the repair of an instrument or inside line.

The case is made of leather which folds in such a manner that when closed it is impossible for the contents to become dislodged from the case. The inspector's pocket kit is shown in figure 8-22.

Marking

The carrying case has stamped into the leather the following:

INSPECTOR'S POCKET KIT,
U.S. SIGNAL CORPS.
No._______

Equipment

1 screwdriver and skinning knife, combined with safety spring, Empire, No.372.
1 scissors, electrician's, 5-inch, nickeled, J.Wiss & Sons.
1 pliers, 5-inch, side cutting, nickeled, pair, U.D.F.&T. Co., No.1250 or P.S. & W., No.1240.
1 file, bastard, 3-inch, half round, with handle
1 tweezers, 4½ inch, nickeled, pair.
1 screw driver 2-inch, Tuck's Giant.
1 rule, 2-foot, narrow, fourfold, boxwood, brass bound, C.S. Co., No. 62½

Electric drills, spring hammers, and other special tools are kept in stock at Signal Corps general supply depots, and are issued to construction parties when the magnitude of the work involved warrants such action. These special tools should be invariably returned to supply depot from which received, upon completion of the work for which they are issued, unless instructions for different action are issued by the Chief Signal Officer of the Army.

Author's note: *The manufacturers listed above by the War Department are not consistent but vary from kit to kit.*

EMPIRE WINSTED CT.

Empire Cutlery made the above knife, which is known as an electrician's knife. The screwdriver blade on these knives is locked open by a backspring lock. To release the lock to close the screwdriver blade it is necessary to depress the main blade which, by way of a spur, lifts the backspring for the screwdriver blade and allows it to be closed.

all similar knives

OAL: 3⅝ inches
Liners: brass
Bolsters: nickel silver

Empire Knife Company began in 1856 and was at its peak when it made the knife shown at the top of the photo. E.C. Simmons Hardware Company of St. Louis, Missouri, owned the Walden Knife Company of New York, which made knives for them under the name of Keen Kutter. The knife in the middle of the photo below has the Keen Kutter logo. The bottom knife was made by the W.R. Case & Son Cutlery Company of Bradford, Pennsylvania. Case is a very old and respected American knife maker of great fame and they have produced probably the most collected knives in the world.

EMPIRE
WINSTED
CT.

E.C. SIMMONS KEEN KUTTER ST LOUIS MO

CASE TESTED XX

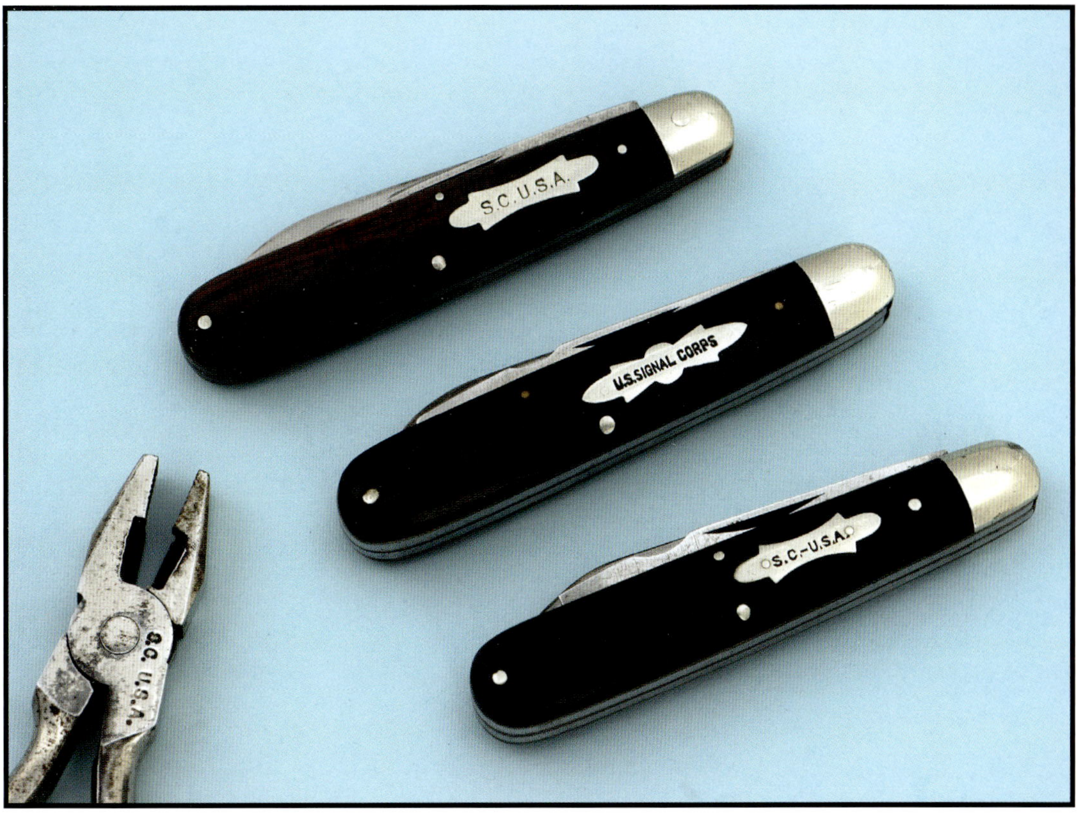

The top knife with the shield marked, "**S.C. U.S.A**." (Signal Corps United States Army), is by Camillus Cutlery Company of New York. The middle knife with the bow tie shield marked, "**U.S. SIGNAL CORPS**" is by Ulster Knife Company of New York. The bottom knife bears the Keen Kutter mark and has a shield marked, "**U.S. SIGNAL CORPS**." Miller Brothers Cutlery Company also made this pattern for the Army Signal Corps. It bears a shield marked "**S.C.-U.S.A.**"

CAMILLUS
CUTLERY CO.
CAMILLUS, N.Y.
U.S.A.

ULSTER
KNIFE CO
N.Y.

E.C. SIMMONS
KEEN
KUTTER
ST LOUIS
MO

An electrician's knife by Ulster Knife Co. with a very unusual shield marked "U.S.N." This is the only one of these I've seen.

ULSTER
KNIFE Co.
U.S.A.

This Signal Corps knife was made by W. R. Case & Son Cutlery Company and is marked with the typical Case logo. The escutcheon plate bears the "S.C.U.S.A." mark but also displays the TL-29 designation. I believe this is the earliest use of this identification of a knife as Tool for Lineman-29.

CASE TESTED XX

TL-29

VALLEY FORGE CUTLERY CO.

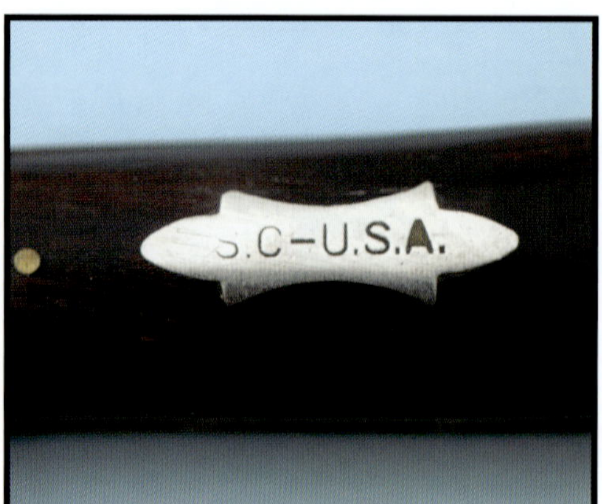

This Signal Corps electrician's knife has been modified with the removal of the screwdriver blade, backspring, and center liner. It is, however, the only example of a Signal Corps knife by this maker that I have been able to find.

Although considerably more robust than the early Signal Corps knives this TL-29 size knife has a screwdriver locking mechanism like that found on the early Signal Corps knives. It also has a serial numbered oversize shield not found on later electrician's knives. The tang is stamped Challenge Cutlery Company and this knife was made prior to 1928 and most likely after World War I. I believe this knife is a transition piece between the Signal Corps knives and the TL-29s.

CHALLENGE
CUTLERY.Co

OAL: 3¾ inches
Liners: brass
Bolsters: nickel silver

H. & J.W. King

H. & J. W. King is the mark on the tang of this early transition knife that displays features of the Signal Corps knives of the First World War as well as those of the TL-29s of later periods. The date, 1856, is part of the maker's trademark and does not represent the date of manufacture. The bail on this knife is made of brass. The shield is marked "U.S. Army" over "P16," which is probably a control number. The knife has the locking screwdriver released by depressing the main blade.

OAL: 3¾ inches
Liners: brass
Bolsters: nickel silver

This is a doctor's knife from the World War I period. In fact, each of these knives is dated. This particular knife is dated 10-22-14. Others have been observed with the date 11-8-12. It was made in Germany and imported by Adolf Kastor, the owner of Camillus Cutlery Company. This was probably one of the last to be imported as the war was soon to end trade with Germany.

main blade obverse

Med. Dept.
U.S.A.

saw blade

KASTOR
10-22-14

main blade reverse

A. KASTOR
& BROS
GERMANY

OAL: 4⅛ inches
Liners: brass
Bolsters: nickel silver

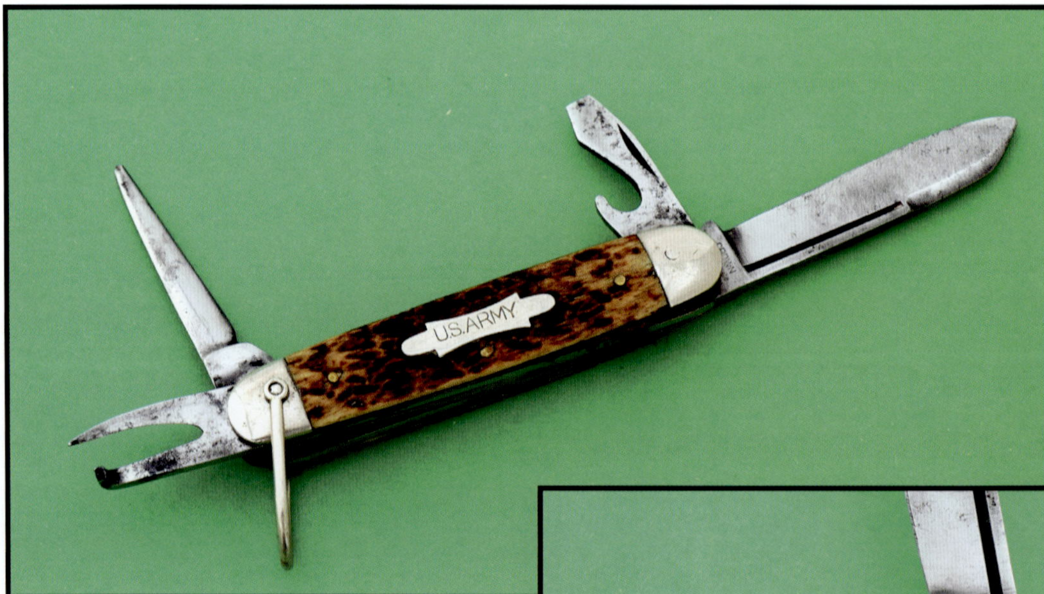

Crown Cutlery Company
New York, U.S.A.

Perhaps the first of the Army's engineer's knives. Crown Cutlery Company in New York was in business only until 1930. The shield and its marking are unique and I can only conjecture that it was made for and issued by the U.S. Army.

WORLD WAR II

This war changed the world, and nothing that existed before was ever the same again. And this was true here in the United States as it was everywhere. Most of us have grown up with wars ... Korea, Vietnam, Panama, the Gulf War, Afghanistan. These we watch on television and they seldom affect the way we live. World War II was different. You couldn't go to the industrial supply store and buy brass. No big thing you say because not many of us do that today. But what about sugar, gasoline, and tires for your car? These items were rationed because they were needed for the war effort. You were immersed in the war. Your daily way of living was involved in the war. What you cooked, what you ate, what you planted and grew, what you made was all affected by the war. The war was part of your everyday life and the decisions you made.

With the start of the war the military found they were in short supply of knives and began programs to accept those knives donated by citizens. Sportsmen, outdoorsmen, and hunters turned in fixed blade knives by the thousands for distribution to servicemen bound for foreign fronts. Industry began to gear up to fill the demand of the military for all sorts of knives. Over the next few years millions upon millions of knives were produced and put into service. Camillus, Case, Imperial, Ka-Bar, Schrade, Ulster, Utica, and Western are but a few of the cutlery companies that geared up production for the war effort, and they filled the need despite a diminishing supply of qualified workers. The knives that are shown on the following pages represent their effort. Over the war years the quality of manufacturing declined. To a large extent this was due to practicality of producing large quantities of knives rather than focusing on a high quality fit and finish. The standard of knife making was lowered during the war and was never to return to its previous heights. After the war, society had changed, and the demand for knives had fallen off as more and more people moved to the big cities. People no longer found a need to carry a knife in their pocket. There were clippers for fingernails, erasers for mistakes, letter openers for mail, scissors for string, and a thousand other nifty tools that once were the purview of the pocket knife. However, the story that follows is about the great knives that were produced by Americans because other Americans needed them in their effort to preserve freedom for the future.

Camillus Cutlery Company of Camillus, New York probably produced more knives during the war than any other single maker of knives. They made the Mark 1, the Mark 2, the M3 trench knife, the M4 bayonet, the USMC fighting/utility knife, the Marine Raider Stiletto, pilot's knife, and millions of pocket knives. The knife shown below has the four-line marking shown at the left, a marking that is common on World War II pocket knives. This knife was called the Engineer's Pocket Knife. It is not unusual, even today, to find one of these in unused condition.

CAMILLUS
CUTLERY CO.
CAMILLUS.N.Y.
U.S.A.

PAL BLADE CO.
MADE
IN U.S.A.

CAMILLUS
CUTLERY CO.
CAMILLUS.N.Y.
U.S.A.

OAL: 3⅝ inches
Liners: brass or steel
Bolsters: nickel silver or steel

The engineer's knife was made by a number of companies including Camillus, Ulster, Imperial, Pal, and Western. Most had steel liners and bolsters as brass was a strategic material and tightly controlled and it was not available for knife making. However, some knives made early in the war do have brass liners. Many of the knives were marked only "Made in USA" and these are probably the product of a joint effort by Imperial Knife Company and Ulster Knife Company, which produced more than half of the knives purchased by the military. These Engineer's knives are 3⅝ inches long when closed and are made with all-steel fittings. Examples of the Camillus knife with brass and nickel silver fittings have been noted.

The military name for these knives was "Knife, Pocket, Engineer, Four Blade, With Clevis." The Army specification design number was 17-170. The Navy called it a "Four Blade, Utility, Jack Knife" and their specification number was 41-J-4. Millions of these knives were produced for the military. Camillus made more than 3.2 million for the Army and another 1.7 million for the Navy and today it is not unusual to find Camillus-marked engineer's knives. Finding those that are marked by other companies is a bit more difficult. The knife shown below is by Western Cutlery of Boulder, Colorado. It is 3⅝ inches in length and has brass liners and nickel silver bolsters.

WESTERN
BOULDER, COLO.

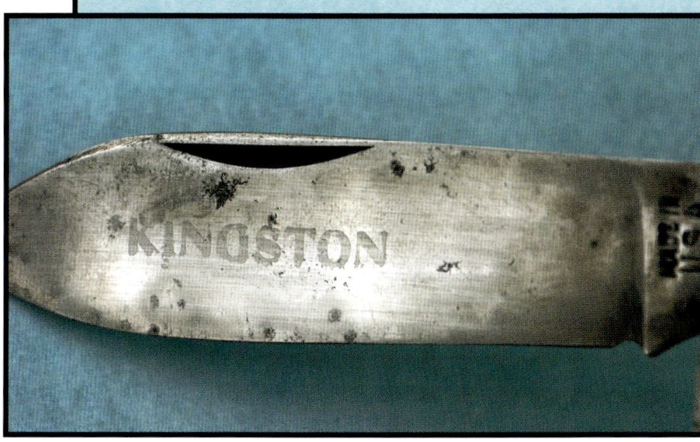

KINGSTON

This engineer's knife is tang-marked "MADE IN/USA." The blade center is etched KINGSTON. Kingston was the name given to a wartime joint venture by Imperial Knife Co. and Ulster Knife Co. They produced thousands of knives for the military under this banner but one seldom finds the Kingston name on their products. One exception to this is the all-metal four blade engineer's knife that later replaced all the bone handled knives. Oftentimes one finds the bail on these knives stamped with the Kingston name.

MADE IN
USA

**MADE IN
U.S.A.**

Collectors run into many engineer's knives marked "Made in USA" and a number of variations have been observed. All of the knives in the above photos are "Made in USA" marked. Note the way the bone was jigged on the top knife and the variety of can openers shown on this page. Can openers were all but standardized in 1944 with the appearance of the safety can opener. All of these knives are $3\frac{5}{8}$ inches long when closed excluding the bail. All are made with steel fittings.

It is not uncommon to encounter Engineer's knives by Camillus that have a shield marked "U.S.A." Marks from other branches of service are also observed from time to time. Shown below are two Camillus knives with the "U.S.M.C." mark. The Engineer's knife is marked with the Camillus three line mark, which is generally considered a late or post-war mark. The electrician's knife below has the four line mark but is unusual in that it does not have the bail typically associated with the TL-29. Both knives are $3\frac{5}{8}$ inches closed and are made with brass liners. The engineer knife has steel bolsters while the TL-29 has bolsters of nickel silver.

CAMILLUS
CUTLERY CO.
CAMILLUS.N.Y.
U.S.A.

CAMILLUS
NEW YORK
U S A

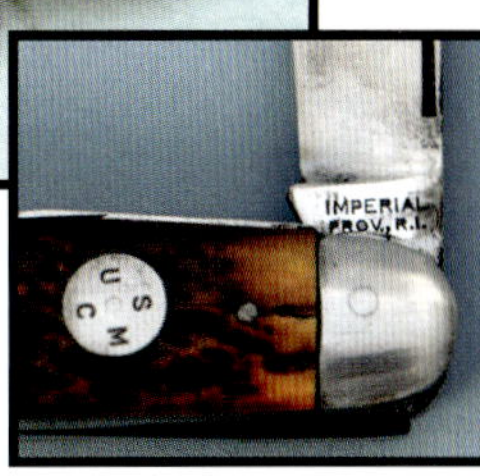

Other knives with the "U.S.M.C." branch of service markings are shown above. The top knife is the Camillus shown on the previous page. The center knife was made by Imperial Knife Company. Detail of this knife is shown in the inset. Note the long pull and the upper bevel on the knife by Imperial. The bottom knife is marked "Made in USA."

CAMILLUS
NEW YORK
U S A

IMPERIAL
PROV, R.I.

MADE IN
USA

M.D.- U.S.N.

The knife at right was made by Utica Cutlery Company of Utica, New York. In 1937 they adopted "Kutmaster" as their trademark. The left knife is by Camillus and bears the four line mark. Both knives are four blade engineer's knives marked for the Medical Department of the U.S. Navy. These knives are 3⅝ inches when closed not including the bail.

KUTMASTER
UTICA, N.Y.
MADE IN U.S.A.

CAMILLUS
CUTLERY CO.
CAMILLUS.N.Y.
U.S.A.

Imperial

This engineer's knife has the blade tang of the main blade marked MD-USN. The screwdriver blade is marked Imperial/Prov, R.I. The knife exhibits standard mid-WWII construction materials but is unique in its tang marking.

OAL: 3¾ inches
Liners: steel
Bolsters: steel

UTICA
CUTLERY CO.
UTICA, N.Y.
MADE IN U.S.A.

This knife displays a level of quality not expected to be found on a wartime knife. The fit and finish shown on this knife is certainly up to pre-war standards but with its steel liners and bolsters, this knife was obviously made during the middle of the war.

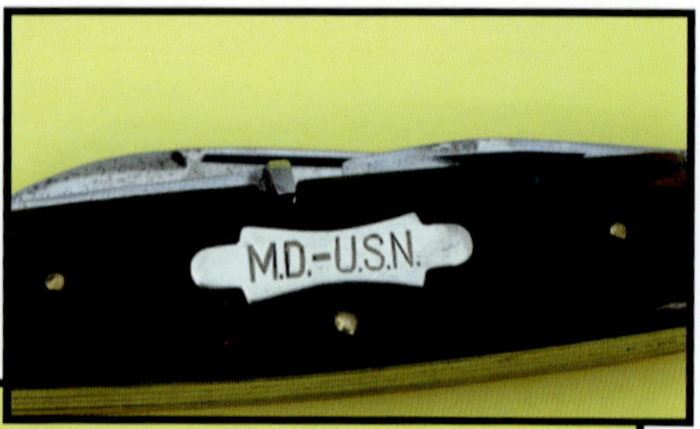

KA-BAR
U.S.

The knife shown below is marked as shown at the left. The KA-BAR mark is famous for its Marine Corps Fighting and Utility Knife of World War II. Although there were several companies that made the Fighting/Utility Knife the name KA-BAR came to represent all knives of that type. The four blade knife below is apparently United States military but it is not the engineer knife because of the blade configuration. It is an interesting item and the author would like to hear from anyone who might be able to shed some light on the subject.

OAL: 3⅝ inches
Liners: brass
Bolsters: nickel silver

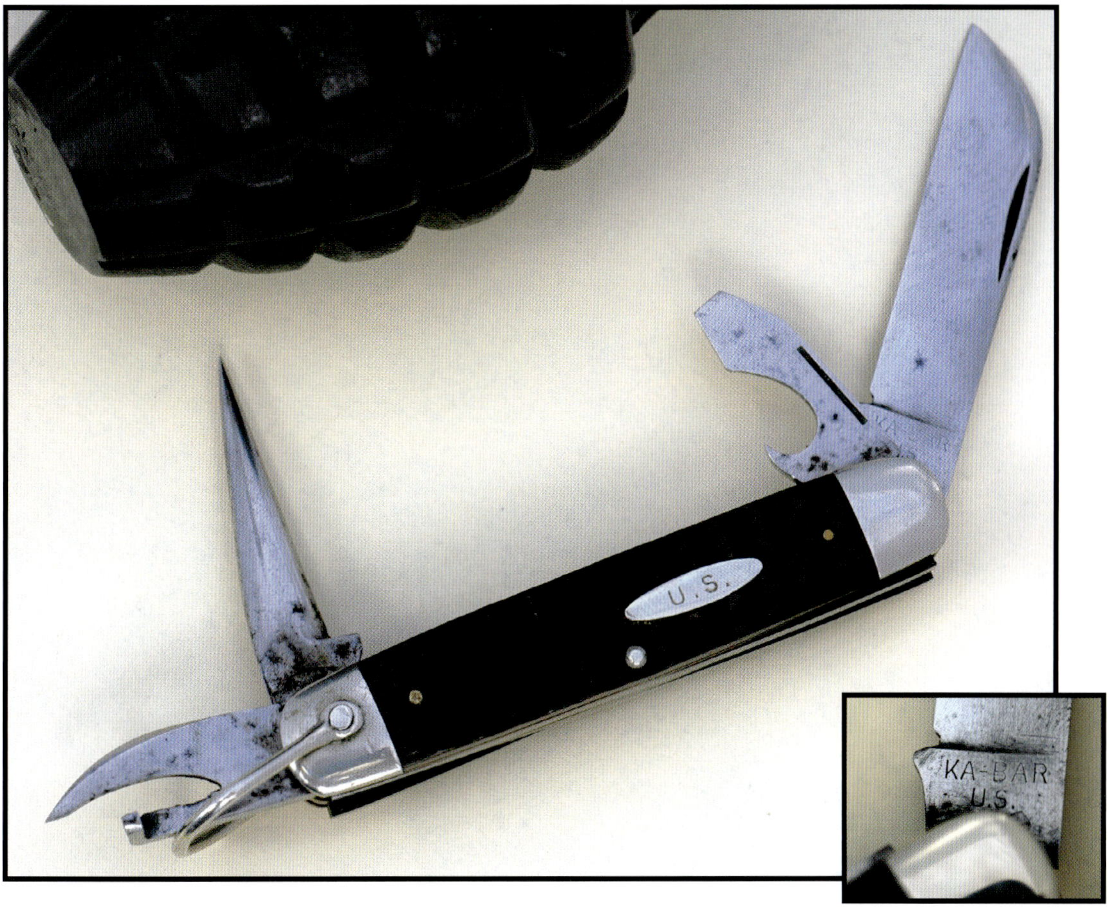

In addition to the four blade engineer's knife there was a three blade utility knife. This Navy-marked knife is by Camillus. These knives have steel bolsters but do have brass liners. This knife is 3⅝ inches closed.

CAMILLUS
NEW YORK
U S A

U.S.N.

OAL: 3⅝ inches
Liners: brass
Bolsters: nickel silver

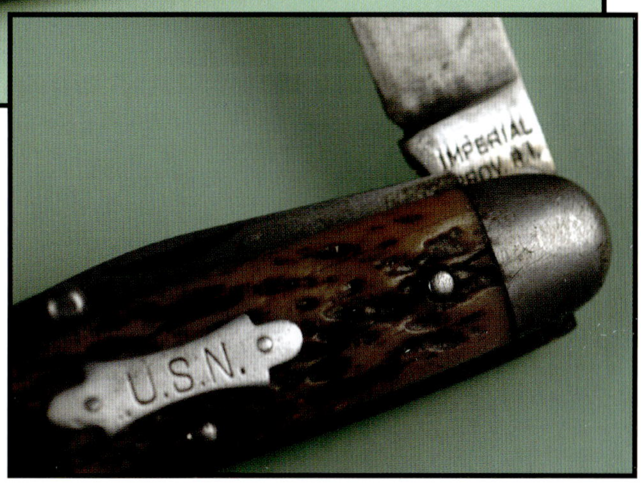

Seldom encountered is the three blade engineer's knife by Imperial with shield marked "U.S.N." Typically you find many more of the same knife by Camillus.

OAL: $3\frac{11}{16}$ inches
Liners: steel
Bolsters: steel

IMPERIAL
PROV, R.I.

This three-blade pocket knife is by Imperial. It is the same pattern as shown on the previous page and was probably made for the for the same purpose but lacked the decorative shield. The knife is $3\frac{11}{16}$ inches closed excluding the bail and is of all-steel construction.

OAL: $3\frac{11}{16}$ inches
Liners: steel
Bolsters: steel

CAMILLUS
CUTLERY CO.
CAMILLUS.N.Y.
U.S.A.

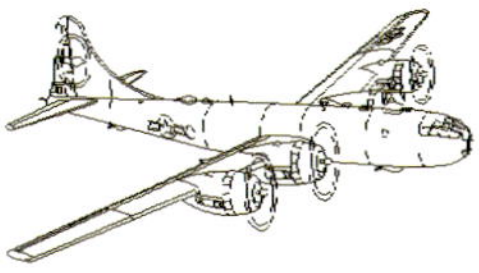

This Army Air Corps pilot's knife is by Camillus and bears the mark at the left. Camillus Cutlery Company made about a million of these knives for the Air Corps.

OAL: 3⅝ inches
Liners: steel
Bolsters: steel

UTICA
CUTLERY CO.
UTICA, N.Y.

CAMILLUS
CUTLERY CO.
CAMILLUS.N.Y.
U.S.A.

The pilot's knife on the left is by Utica Cutlery Company, which later used the Kutmaster trademark. The knife at the right is by Camillus. Both knives are $3\frac{5}{8}$inches in length when closed. The Utica knife has brass and nickel silver fittings while the Camillus knife has fittings of steel.

UTICA
CUTLERY CO.
UTICA, N.Y.

These three pilot's knives all have the same handle and shield configuration but each knife has a different tang stamp. All three knives were made by Utica Cutlery Company.

OAL: $3\frac{31}{50}$ inches
Liners: steel (bottom knife has a brass center liner)
Bolsters: steel

MADE IN U.S.A.

OAL: 3⅜ inches
Liners: steel
Bolsters: steel

This three-blade pocket knife is marked as shown on the blade. It is a small 3⅜ inches when closed and has liners and bolsters of steel. Its origin and purpose are unknown.

PAL BLADE CO.
MADE
IN U.S.A.

CAMILLUS
CUTLERY CO.
CAMILLUS.N.Y.
U.S.A.

IMPERIAL
PROV, R.I.

UTICA
CUTLERY CO.
UTICA, N.Y.
MADE IN U.S.A.

MADE IN
U.S.A.

U.S.A.

The knives on the preceding page and those shown below are Navy general utility knives. These knives were made by Camillus, Utica, Pal, Imperial, and probably by Kingston which was the name of the joint operation of Ulster and Imperial. The knives by Kingston are marked "Made in USA." The knife on the left is probably by Kingston while the knife on the right is by Utica. These knives are 3⅜ inches in length and some are found with steel fittings while others are found with brass and nickel silver.

MADE IN U.S.A.

UTICA
CUTLERY CO.
UTICA, N.Y.
MADE IN U.S.A.

PAL BLADE CO.
MADE
IN U.S.A.

This Navy general utility pocket knife was made by Pal Blade Company. The Pal Blade Company purchased the cutlery division of Remington Arms Company in 1940. This knife has all-steel liners and bolsters. Others have been observed with brass liners and nickel silver bolsters.

all knives in this category

OAL: 3⅜ inches
Liners: brass or steel
Bolsters: nickel silver or steel

CAMILLUS
CUTLERY CO.
CAMILLUS.N.Y.
U.S.A.

IMPERIAL
PROV, R.I.

The knife on the left was made by Camillus Cutlery Company and the one on the right was made by Imperial. The small blade on the Imperial made knives is marked "U.S.A."

This two-blade pocket knife by W. R. Case & Son Cutlery Company was made for Army Air Corps pilots during the war.

OAL: 3¾ inches
Liners: steel
Bolsters: steel

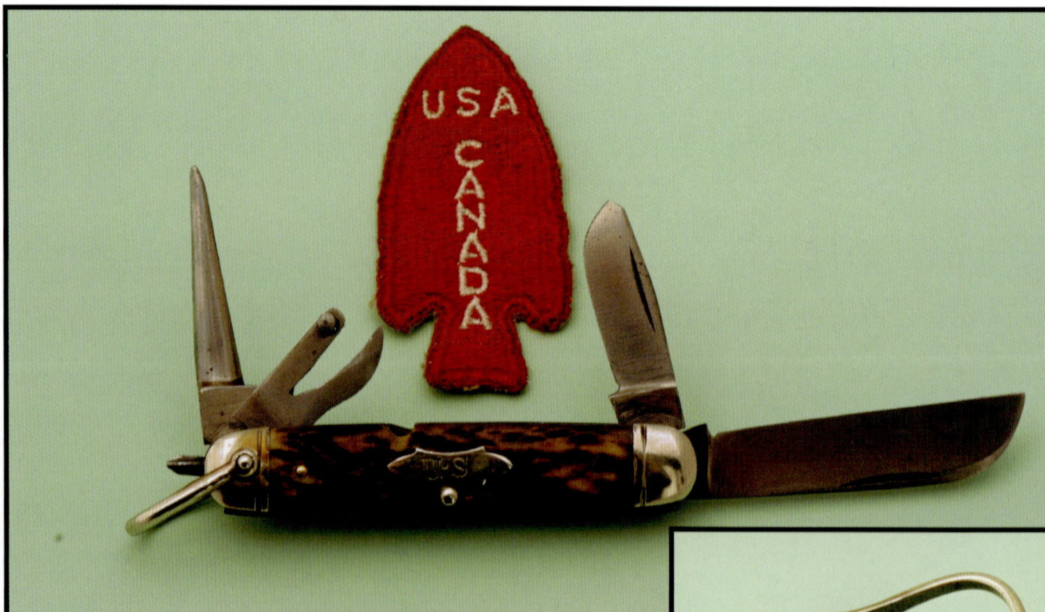

ULSTER MOUNTAIN POCKET KNIFE

The Ulster Mountain Pocket Knife was provided to troops of the First Special Service Force and others. This piece came out of the Ulster factory collection sold around 2012. It was a prototype for the variations shown on the next few pages. The Phillips head screwdriver on this model was sandwiched between two center liners and held in place using the pivot pin at that end of the knife.

ULSTER
KNIFE Co.
U.S.A.

These knives were produced for the mountain and ski troops. The bail on these knives had a Phillips screwdriver, which was necessary for adjusting ski attachments. The knives are relatively small for a four-blade knife.

OAL: 3⅝ inches
Liners: brass
Bolsters: nickel silver

The First Special Service Force was a unit composed of American and Canadian forces. They were very effective fighters in Europe and became known as the Devil's Brigade. They were also the forerunners of today's Special Forces, the Green Berets. Part of their training included mountaineering and the Mountain Pocket Knives shown on the previous page and below were useful in those operations. The Phillips screwdriver on this model was added as a separate blade. The knife shown at the top is of all-steel construction while the knife at the bottom has brass outside liners.

ULSTER
KNIFE Co.
U.S.A.

OAL: 3⅜ inches
Liners: brass or steel
Bolsters: nickel silver or steel

obverse

PRESTO
PAT.PEND-G

reverse

G.SCHRADE
B'PORT.CT

American paratroopers were supplied with automatic knives that could be opened with one hand so in an emergency if it became necessary to cut tangled shroud lines. The first of these knives were made with bone handles. The mark side of the knife bears the mark shown at the top left while the pile side bears the mark shown below it.

OAL: 4⅛ inches
Liners: steel
Bolsters: steel

obverse

SCHRADE
CUT.CO.
WALDEN,N.Y.

reverse

U.S.PATS.
DEC.21,09.
SEPT.13,10.
JUNE 6,16.

The knife shown above was made by the Schrade Cutlery Company of Walden, New York. This company was formed in 1904 by three brothers: Jacob Louis, George, and William Schrade. In 1910, George left the firm and later started the company that made the knife shown on the previous page. The knife above is 4¼ inches long and has steel liners and bolsters. The mark side of the knife is imprinted with the logo shown at the upper left and the pile side of the knife bears the mark shown at the lower left.

OAL: 4¼ inches
Liners: steel
Bolsters: steel

obverse

PRESTO
PAT.JAN.30-40
MADE IN U.S.A.

reverse

GEO.SCHRADE
KNIFE CO.INC.
B'P'T.CONN.

This switchblade is shown with its original wrapper and lanyard and remains as fresh as the day it was issued. The obverse side of the knife bears the mark shown at the upper left and the reverse side is marked as shown at the lower left.

OAL: 4⅛ inches
Liners: steel
Bolsters: steel

obverse

PRESTO
PAT.JAN.30-40
MADE IN U.S.A.

reverse

GEO.SCHRADE
KNIFE CO.INC.
B'P'T.CONN.

This knife with the hawkbill or pruning style blade probably served well as a line cutter. The mark side of the blade bears the mark shown at the upper left while the pile side is marked as shown at the upper right.

OAL: 4⅛ inches
Liners: steel
Bolsters: steel

obverse

PRESTO
MADE IN U.S.A.

reverse

GEO.SCHRADE
BRIDGEPORT
CONN.

As the war progressed manufacturing techniques designed to reduce costs were tested. A number of experimental bayonets were tested and some firearms like the M3 sub-machine gun, (better known as the "grease gun"), were made of stamped sheet metal parts. Likewise, the two automatic knives shown above have stamped steel handles resulting in a knife made totally of steel. The small knife on the left is 4⅛ inches long and the knife on the right measures about five inches in length. Markings on both knives are identical with the obverse side marked as shown at the upper left and the reverse side marked as shown at the lower left.

Both of these knives were made in England for the British Special Operations Executive (SOE). They were made with a tire slasher blade on one end and a clip point lockback blade on the other. They have some type of plastic handles that show considerable shrinkage. Both knives are unmarked.

This British gravity knife was made by George Ibberson and is so marked on the blade. The knife also possesses a number under the spike. These were issued to paratrooper and commando units. Unmarked versions were issued to SOE and OSS personnel. The knife is nearly a direct copy of the German paratrooper knife but rather than the wood handles the British version has checkered plastic scales. Frederick Stephens in his book *Fighting Knives* suggests that more than 100,000 of these were made for the war effort. While this may be the case these are extremely rare today and no one knows what has happened to the bulk of the supply.

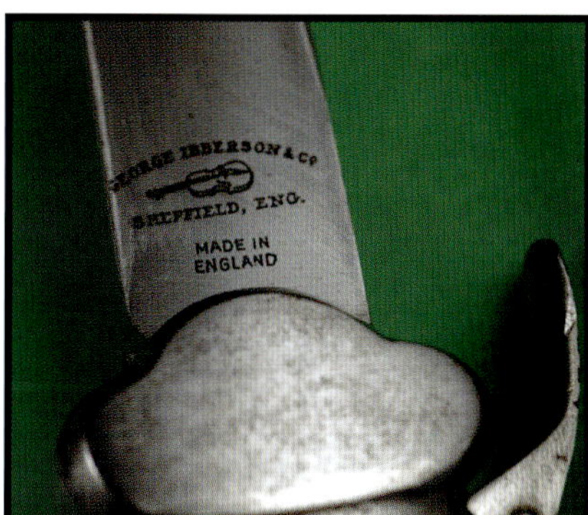

OAL: 6¼ inches
Liners: steel
Bolsters: steel

There are two versions of this British knife. One is marked A (crown) M for Air Ministry and the other is unmarked. Most of the unmarked versions are now marked Made in England because, I believe, they were purchased as military surplus in England and brought into the United States for resale thus requiring the country of origin mark. There are two blades, a spear point main blade and a very heavy spear point spike. The use of this second blade is unknown. These knives were used in survival kits like the one shown below.

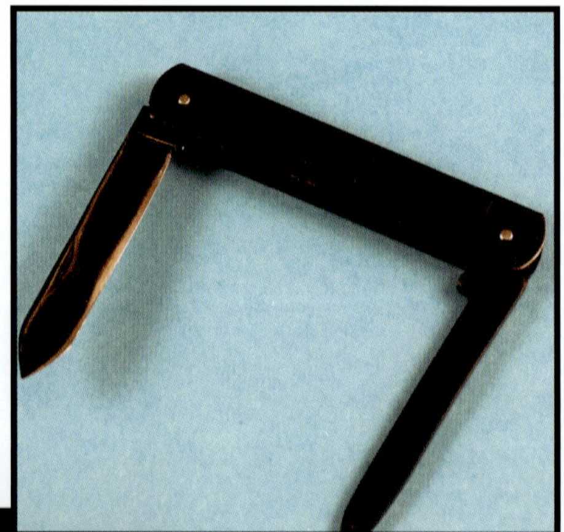

OAL: 4¾ inches
steel stamping

This knife is reportedly a prototype for the SOE and OSS escape knives shown on the following pages. The knife is made by Joseph Rodgers and has steel liners. The scales are made of some type of black plastic. The center liner provides a screwdriver as it extends past the end of the knife. The knife is 4¼ inches long when closed.

Fig.8

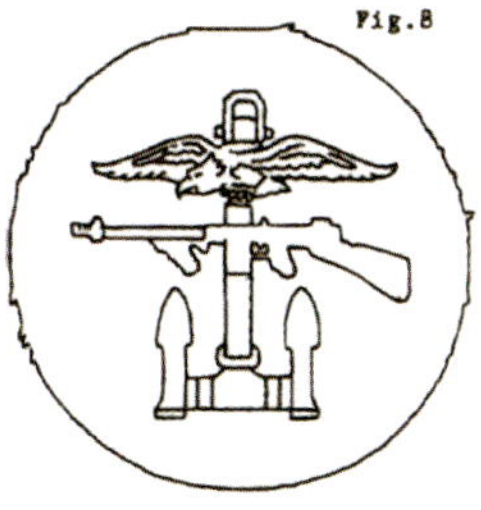

During the war, the British developed an escape knife that was to be used by those who might need a small, concealable tool useful for escape and evasion purposes. British Special Operations, Executive (SOE) and the Office of Strategic Services (OSS) often shared equipment and it is likely that some of these were used by U.S. servicemen in the European theater of operations. The tool, originally made during the Victorian period, was developed as an escape tool during the war by the addition of a screwdriver tip and three saw blades. Both the early Victorian model and a prototype are marked Joseph Rodgers & Sons.

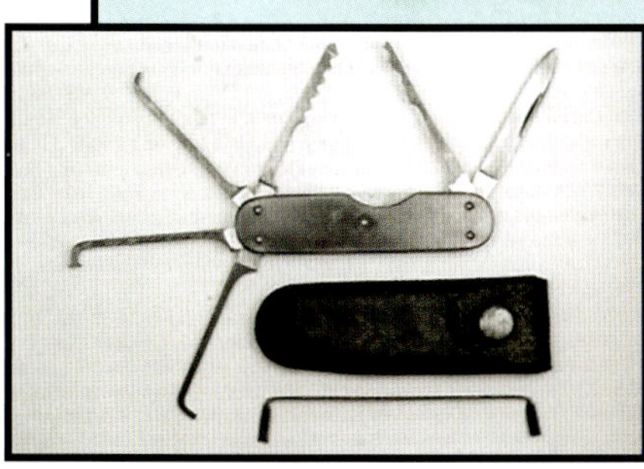

The escape knife shown above is the issue knife and is totally blackened and completely unmarked. It is of all-steel construction and is 5⅜ inches long and weighs slightly more than half a pound. The knife shown in the inset is an OSS issue lockpick knife. It is 2⅞ inches in length and has six "blades." It was supplied in a leather pouch with a torsion bar, which was used with the picks to open a lock. The knife is completely unmarked. Post-war knives made and supplied to the Central Intelligence Agency (CIA) were marked "SCHRADE/ WALDEN/NY USA." The original OSS knives are quite rare and only three or four are known to exist in collections today. Inset from Dr. John W. Brunner, author of *OSS Weapons.*

The OSS lockpick knife is of all-metal construction. It measures just 2⅞ inches long when closed. It came in a small leather pouch along with a torsion tool that is used in the lockpicking process. The knives are unmarked but were made by Schrade Cutlery Company in Walden, New York. Dr. John Brunner in his book *OSS Weapons* states that the OSS placed an order for 1,000 knives and by the war's end only fifty-six remained in inventory. The knives are very fragile and the blades are easily broken. Less than half a dozen are known to be in collections.

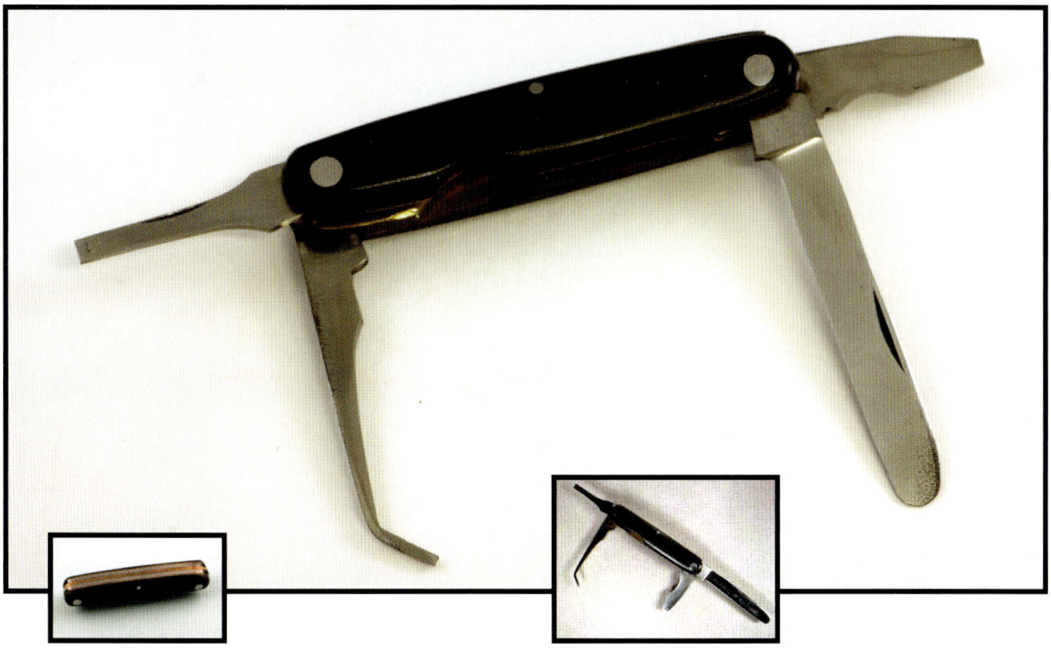

This very odd four blade knife has black Bakelite handles. The blades are unique in appearance and design. The center spacer is quite thick and is milled out to allow the blade with the crooked end to have a place to register when closed. On some specimens, the main blade is marked "Property of U.S. Govt." Others are unmarked. The main blade has a rounded point but a sharpened edge. The other two blades are screwdrivers. My guess is that the knife is some sort of adjusting tool. Hopefully, some additional information will be forthcoming. There is no makers mark on the knife.

OAL: 3⅜ inches
Liners: brass
No bolsters

TL-29

One of the most notable pocket knives to come out of World War II is the TL-29 (Tool for Lineman, 29). This is an electrician's knife with a spearpoint main blade and a locking screwdriver blade. The knife was originally carried in a leather pouch like the one shown below. The web carrier marked, "POUCH CS-35A" was provided during the War when leather was in short supply. The leather one is marked, "POUCH TYPE CS-34/SIGNAL CORPS U.S. ARMY." The pouch also carried the electrician's pliers, the TL-13A.

For World War II the inspector's pocket kit was brought back into service and renamed Tool Equipment TE-5. Tool quality was reduced somewhat but the package was still a great kit. Unlike the earlier kits, except for the electrician's knife, none of the tools carried the Signal Corps markings.

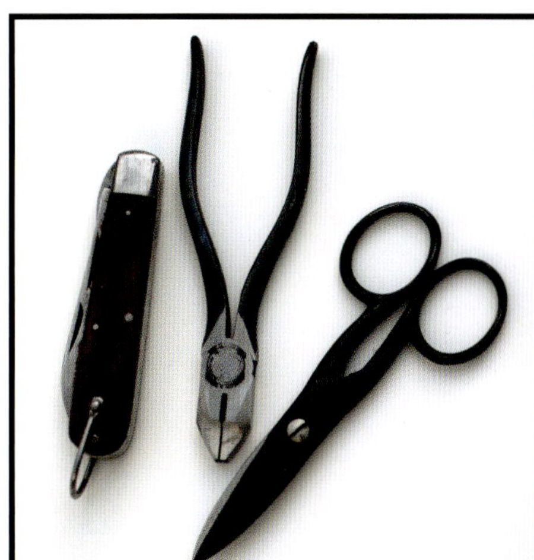

PAL BLADE CO.
MADE
IN U.S.A.

The screwdriver blade on these TL-29s was locked open when a portion of the liner would spring into its path to prevent it from closing. Closing instructions, "To Release Push Center Lock to Left," were etched on the blade. This early bone-handled electrician's knife also has "U.S. ARMY" etched on the screwdriver blade. The main blade is marked as shown above.

OAL: 3½ inches
Liners: steel
Bolsters: steel

PAL BLADE CO.
MADE
IN U.S.A.

U.S.ARMY

Few of the TL-29s were bone-handled and the only ones which have been marked appear to have been made by Pal Blade Company. The knife shown below has a main blade whose pile side was stamped as above on the right. The obverse side is marked as above on the left. The knife is 3½ inches long closed and all fittings are steel but those with brass liners have been observed.

Here are two more TL-29s by Pal. The wood is stamped "TL-29" on the upper knife and "U.S. ARMY" on the lower knife. Both blades are marked as shown at the right.

OAL: 3½ inches
Liners: brass and steel
Bolsters: steel

PAL BLADE CO.
MADE
IN U.S.A.

CAMILLUS
CUTLERY CO.
CAMILLUS.N.Y.
U.S.A.

Camillus Cutlery Company of Camillus, New York, made millions of pocket knives for the military during World War II. This was one of the first TL-29s they made during the period. This early wartime model is unusual in that it has jigged bone handles.

OAL: 3¾ inches
Liners: brass
Bolsters: steel

The knife shown below with the escutcheon plate marked "TL-29" was made by Camillus Cutlery Company. These knives were a bit longer than those made by the Pal Blade Company. Both knives in the inset were made by Camillus but note the unusual shield on the bottom knife.

CAMILLUS
CUTLERY CO.
CAMILLUS.N.Y.
U.S.A.

OAL: 3¾ inches
Liners: brass
Bolsters: steel

Most of the military World War II era electrician's knives were marked by stamping the wooden handle scales or an escutcheon plate with "TL-29." However, a variety of other marks and marking methods have been found. Some of the stamped marks have been filled with paint, usually gold by Camillus and Utica while Kutmaster used white. Camillus made one version that was marked only with a painted on "TL-29" in black paint. Both Kutmaster and Camillus stamped the handle on one version with the number "R41 - K - 455." The Kutmaster version is shown below. This number identifies it as a knife belonging to the Navy Bureau of Aeronautics. Electrician's knives are also found with the branch of service markings. The photo below depicts the "U.S.N." marked knife by Ulster, Camillus, and Kutmaster.

ULSTER
KNIFE Co.
U.S.A.

CAMILLUS
CUTLERY CO.
CAMILLUS.N.Y.
U.S.A.

KUTMASTER
UTICA, N.Y.
MADE IN U.S.A.

R = Aviation Supplies
41 = Stock class, hand tools
K = Knife
455 = Stock number

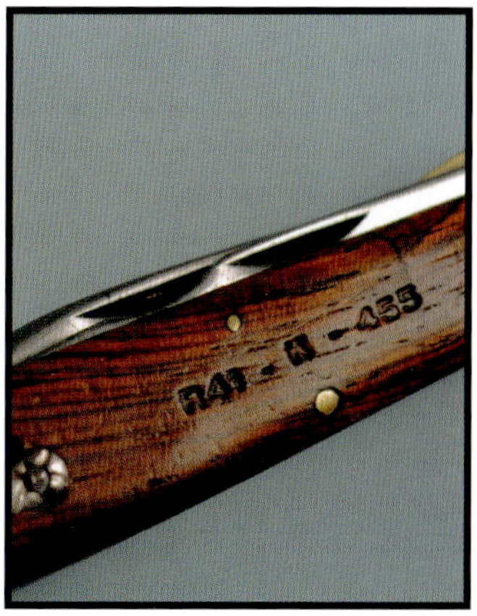

Another four line Camillus marked knife with wooden scales simply stamped "U.S.A." TL-29s with this mark are seldom seen.

The two TL-29s above are by Camillus Cutlery Company. They are 3¾ inches long when closed and both have brass liners and steel bolsters. These knives have also been observed with steel liners and bolsters. The knife on the left has the four line mark and the one on the right has the three line mark.

CAMILLUS
CUTLERY CO.
CAMILLUS.N.Y.
U.S.A.

CAMILLUS
NEW YORK
U S A

CAMILLUS
CUTLERY CO.
CAMILLUS.N.Y.
U.S.A.

This TL-29 by Camillus is typical of the high quality World War II era knives. Blades were nicely shaped, had a good polish, and were nicely etched and marked.

OAL: 3¾ inches
Liners: steel
Bolsters: steel

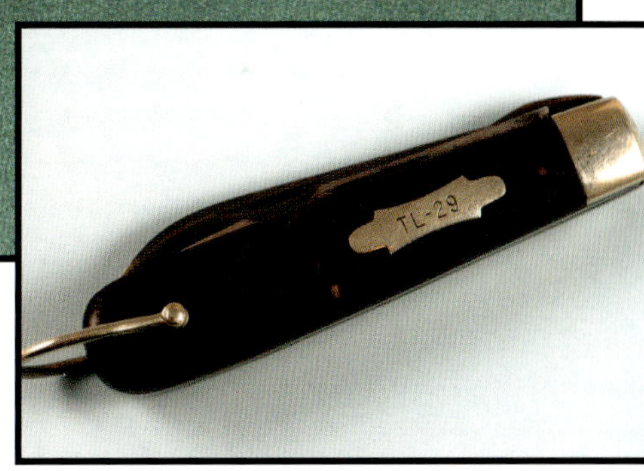

CASE
TESTED XX

This electrician's knife by Case is marked "TL-29" on the wooden handle. The one in the inset is an earlier version and displays the marked metal shield.

OAL: 3¾ inches
Liners: steel
Bolsters: steel

CATTARAUGUS
CUTLERY Co.
LITTLE VALLEY
N.Y.

Cattaraugus Cutlery Company began operations about 1886 and continued until about 1963 when they closed the doors. During the War they made these handsome TL-29s. Both knives are $3\frac{5}{8}$ inches long. The knife at the left has outside liners of brass but the central liner, the locking liner, is steel. The wood handle is stamped "TL-29." The knife at the right has brass liners. The knife is stamped "U.S.N." on the blade. The center liner is marked, "PATENTED/JULY 3, 1906."

There are a number of varieties of the Cattaraugus TL-29 but the one shown above is the most common. Like others, they are marked 21087 on the reverse of the main blade tang. The knife shown below is marked TL-29 on the shield. Very few of these have been observed among collections.

The above knife proudly displays its Marine Corps heritage on the shield while the knife shown below has no shield and lacks the makers model number on the reverse of the tang. It also has a center liner of steel. You will note that some of these Cattaraugus TL-29s are equipped with a bail while others are not.

This is a very unusual TL-29 with no makers mark on the blade tangs and only the Cattaraugus markings on the lock tab of the center liner. I have seen several foreign made and marked TL-29 with the same mark and I believe that Cattaraugus had some right to this type of liner lock and these knife makers are recognizing that. This knife certainly does not have the quality associated with Cattaraugus and it perhaps may have its origins out of the country.

UNION CUT.CO.
OLEAN,N.Y.
U.S.A.

This high quality, fine knife was made by Union Cutlery Company, which started out about the beginning of the twentieth century. In 1911 they established themselves in Olean, New York. They later adopted the name KA-BAR as a trademark for their products, a name that is well respected within the United States military for their cutlery, particularly the Marine Corps Fighting/Utility Knife. This knife is about 3¾ inches long and has a shield marked "TL-29." The liners are brass and bolsters are nickeled silver.

KA-BAR

KA-BAR

This knife shows the family resemblance by looking very similar to the preceding knife marked Union Cutlery Company. It also exhibits a similar high quality.

OAL: 3⅝ inches
Liners: steel & brass
Bolsters: steel

Ka-Bar made these two TL-29s during World War II. They have handles that appear to be made of redwood. One of the handles is stamped "TL-29" and the other "U.S. ARMY." Both knives are marked on the blade as shown below:

KA-BAR
OLEAN,N.Y.

OAL: 3¾ inches
Liners: steel
Bolsters: steel

The knife above was made by Remington Arms Company that, before 1940, maintained a cutlery division. The knife is 3½ inches long and has brass liners and nickel silver bolsters. The main blade is marked as shown at the left and the screwdriver blade is marked as shown at the right. The shield is marked "TL-29."

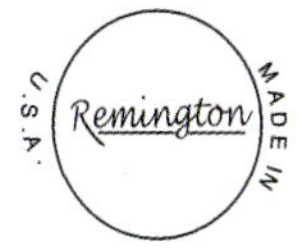

Robeson Cutlery Company of New York made the above TL-29. It is marked as shown at the right and the knife measures 3½ inches long. It has brass liners and steel bolsters and the wooden handle is stamped "TL-29."

ROBESON U.S.A. CUTLERY

These are two early TL-29 marked with "U.S." and an individual serial number on the shield. Otherwise they are identical to later TL-29s. I believe that these were made in the 1930s.

SCHRADE
CUT.CO.
WALDEN,N.Y.

Schrade Cutlery Company started out at the beginning of the twentieth century but sold off their assets and name a few years ago. Almost all of the big names of the 1920s, 1930s and 1940s are gone today. The knives shown below are all marked as above and all measure 3¾ inches in length. The top and bottom knives have brass liners while the middle knife has steel liners. All have steel bolsters. The two bottom knives are identical except for the bails.

This Schrade TL-29 is typical of wartime construction, high quality but containing no brass parts. Note the "TL-29" which is actually "carved" into the wooden handle.

SCHRADE
CUT.CO.
WALDEN,N.Y.

OAL: 3¾ inches
Liners: steel
Bolsters: steel

ULSTER
KNIFE Co.
U.S.A.

This very well-made knife is by Ulster Knife Company of New York and carries a mark that was discontinued at some point during the War. The fit and finish of this knife reflect the high quality of knives bearing the early mark. Later knives show a drop in quality.

OAL: 3¾ inches
Liners: brass
Bolsters: nickel silver

The two knives pictured above were made by Ulster Knife Company of New York. Both knives have brass liners and nickel silver bolsters. The knife on the left is marked with the "TL-29" on the blade and has a somewhat bulbous handle that is quite a bit thicker than those normally found on these knives. The knife at the right has the conventional stamped wooden handle. Both knives measure 3¾ inches closed.

ULSTER
TL29
U.S.A.

ULSTER
KNIFE USA

ULSTER
U.S.A.

Like the knife shown on the preceding page this knife has unmarked, rather robust wood handles. The blade is marked "Ulster/USA" on the obverse and "TL-29" on the reverse.

OAL: 3¾ inches
Liners: brass
Bolsters: nickel silver

Both of these knives are by Ulster Knife Company. The bottom knife bears an earlier mark and shows great quality and workmanship. It is 3½ inches in length and all parts are steel, which would indicate the knife was made in 1943 or 1944. The top knife is a quarter of an inch longer and is fitted with brass liners and nickel silver bolsters. The bottom knife is marked as shown at the left and the top knife is marked as displayed on the right.

ULSTER **KNIFE Co.** **U.S.A.**	ULSTER KNIFE USA

ULSTER
KNIFE Co.
U.S.A.

This TL-29 is by Ulster and it is stamped "U.S.N." on the wooden handle. This knife, which measures 3½ inches bears the mark shown above. It has brass liners and nickel silver bolsters. Ulster also made two knives with unmarked wooden handles. The TL-29 mark was stamped into the blade. See the example earlier in the book. It is marked as shown here:

ULSTER
TL29
U.S.A.

The other is marked as shown below:

obverse	*reverse*
ULSTER U.S.A.	TL-29

This early Kutmaster TL-29 by Utica Cutlery Co. displays the U.S. and serial number on the shield.

KUTMASTER
UTICA, N.Y.
MADE IN U.S.A.

This TL-29 made by Utica Cutlery Company bears their trademark Kutmaster logo at the left.

OAL: 3¾ inches
Liners: brass
Bolsters: nickel silver

KUTMASTER
UTICA, N.Y.
U.S.A.

Both of these Kutmaster (Utica) TL-29s are clearly marked with their tool identification number and their blades are marked as shown at the left.

OAL: 3¾ inches
Liners: brass
Bolsters: nickel silver

TL-29

Another unusual TL-29 is pictured below. At 3¾ inches it is the same length as most of the others but it has a slimmer profile. The only marking on this knife is on the obverse side of the main blade and that mark is shown at left. The knife has steel liners and bolsters and is well-made.

The rope knives of World War II were designed and approved by the Coast Guard as safety equipment carried on ships. The rope knife by Camillus is found both with and without the Coast Guard markings. The mark side of the Camillus knife has the four line mark as shown below. The knife is 4 3/8 inches long and can be found with both brass and steel liners. Bolsters are steel.

Camillus

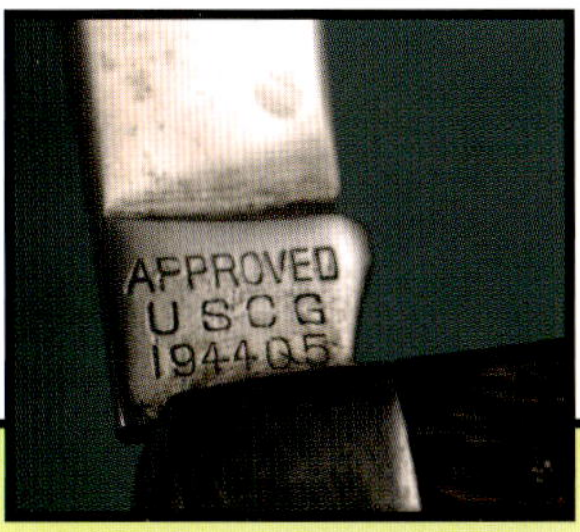

Kutmaster

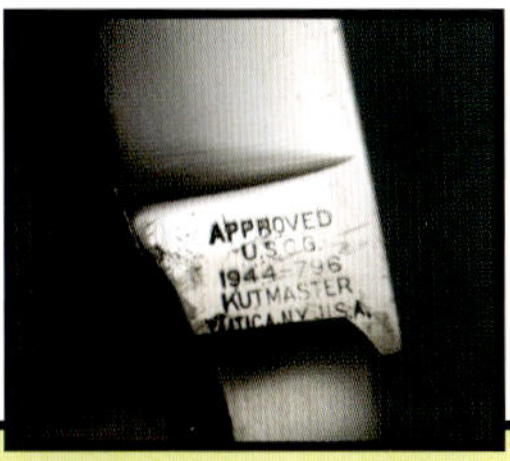

Imperial

obverse

CAMILLUS
CUTLERY CO.
CAMILLUS.N.Y.
U.S.A.

reverse

APPROVED
U S C G
1944 Q5

The Kutmaster and Imperial knives on the preceding page are marked as indicated below.

mark side

APPROVED
U.S.C.G.
1944-796
KUTMASTER
UTICA,N.Y.,U.S.A.

The Kutmaster knife has a length of 4¼ inches and has brass liners and steel bolsters.

mark side	*pile side*
IMPERIAL PROV, R.I.	APP'VD 1944 U.S.C.G. #85C

The Imperial knife has a length of 4 inches and has brass liners and steel bolsters.

Like the rope knives of earlier periods, some ended up as surplus and were turned into advertising incentives by an enterprising supplier.

OAL: 5 inches
Liners: steel
Bolsters: steel

All three of these Emergency Fishing Knives are identical except for their markings. Two are marked with the Camillus four line mark, one with a large imprint and one with a small, and the third knife is marked with the "MADE IN U.S.A."

MADE IN
U.S.A.

CAMILLUS
CUTLERY CO.
CAMILLUS.N.Y.
U.S.A.

CAMILLUS
CUTLERY CO.
CAMILLUS.N.Y.
U.S.A.

The World War II Navy Marlin Spike knife by Camillus is quite robust at $5\frac{1}{8}$ inches long, excluding the bail. The handles are wood which has been jigged to simulate bone, and the liners and bolsters are steel. This model is an inch longer than similar knives commonly encountered with plastic handles.

CAMILLUS
CUTLERY CO.
CAMILLUS.N.Y.
U.S.A.

KINGSTON

OAL: 3¾ inches
Liners: brass
Bolsters: n/a

During the latter part of World War II there was an attempt to replace the engineer's knife, the mountain knife, the electrician's knife and the mess kit knife with a single knife that would meet all the needs. It was decided that an all-metal knife could certainly replace the engineer's knife but not the others. The new knife is shown below and the only marking is "KINGSTON," which is found on the bail. Kingston was the name used by the joint effort of Imperial Knife Company and Ulster Knife Company to produce knives for the war effort.

STEVENSON-45

OAL: 3¾ inches
Liners: brass
Bolsters: n/a

These all-metal knives have an especially long bail. This design feature was established in the attempt to replace the mess kit knife. It was thought that the knife could be hung from the mess kit handle and dunked in hot water along with the kit as a means of sterilizing it. This knife is marked on the bail as shown above. Others have been observed that are marked only, "Stevenson." Stevenson Manufacturing Company of Rochester, New York, had a single large contract for these knives in 1945.

KINGSTON

OAL: 3¾ inches
Liners: brass
Bolsters: n/a

Here are another two of the new "KNIFE, POCKET, GENERAL PURPOSE" as they were officially called. These both bear "U.S. MARINE CORPS" marking on the handle scales. The Marine Corps specification called for the screwdriver blade to be at the same end as the main blade. The first knives were produced in that manner as shown by the knife at the top. Later all knives, both Army and Marine Corps, had the same blade configuration as shown by the bottom knife. The bottom knife is by Kingston. Reviewing the military specification for these knives it was noted the overall length is specified at 3⅝ inches. All the knives I measured came in just under 3¾ inches.

SCHRADE
CUT.CO.
WALDEN,N.Y.

OAL: 3⅝ inches
Liners: aluminum
Bolsters: n/a

This interesting knife has jigged aluminum scales, which are deeply struck, "U.S.A." It is also made quite similar to the specifications for the "KNIFE, POCKET, GENERAL PURPOSE." I am not aware, however, that the style of can opener blade on this knife was ever used by any of our military forces. Perhaps some reader can shed some light on the subject.

The large, heavy duty knives shown below were placed in aircraft survival kits. The knife has one large clip point locking blade and a saw blade suitable for cutting metal. The knives were chemically blackened. There are three variations of this knife. Two are marked and one is totally unmarked. The one made by Colonial Cutlery Company of Providence, Rhode Island, has the blade marked as shown below. Also shown below is the mark for the knife made by United Machine Tool Company, Grand Rapids, Michigan.

OAL: 6 inches
Liners: steel
Bolsters: steel

UNITED
G.R. MICH.

COLONIAL
PROV. R.I.

COLONIAL
PROV. R.I.

The knife shown in the above photo is quite similar to the knives on the preceding page except that it is equipped with a bail. It also was issued in the pouch and was used by naval pilots in their survival vests. The only maker of this knife was Colonial Cutlery Company.

OAL: 6 inches
Liners: steel
Bolsters: steel

POST-WAR

Toward the end of World War II is noticed a decline in the quality of pocket knives. It may not have been as much a decline in quality as it was a decline in human skill in the fabrication of the knife. Much of the horn and bone used for handles came from Germany and this ended during the war. Thus, natural materials gave way to plastics and the character of the pocket knife changed forever. The new knives are just as serviceable or perhaps even more so than the older knives but they don't offer the same warmth or feeling when held in the hand. America's romantic association with the pocket knife evaporated as we moved into the atomic age. The pocket knife was no longer a personal tool of an individual but it became just another screwdriver in the tool box. Eventually it even lost its identification. Gone are the branch of service markings along with the TL-29 stampings. An era has closed but a new period has begun.

The era of high-tech knives has, in recent years, spread through society. The knives we carry today are most likely a single blade lockback of some sort. The blade is not high carbon steel but an alloy steel, most likely a high-quality tool stainless steel that can be hardened to a Rockwell C scale hardness around 60. This steel bears a name like AUS8, ATS34, 154CM, A-2, D-2, L-6, E52100, AUS6, 440C, etc. The open and close mechanisms are designed on a computer and produced on computer-controlled machines whose output is so precise that it is measured in 0.001 inches. The idea of a man hammering a steel pin into a mushroom shaped rivet to hold a bone handle scale into place is as far away as a flint napped obsidian knife blade. Oh yes, a certain amount of skill was needed to make the old knives but that skill has no place in the world of high technology. Today's knives easily open with one hand or a flick of the wrist. They clip to your clothing and are ready for super-fast deployment. They need little care or attention and they stay sharp use after use. Well, perhaps there is a bit of dreaming here as not all the military's knives have reached this advanced state yet but some have and the rest are on their way. Anyway, let us maintain the romance of the old with our collecting and embrace the new with what we carry and use.

SCHRADE
WALDEN
NY USA

These automatic knives, designated as the M2, were supplied to paratroopers and pilots in the 1950s. In military terms, the knife was known as KNIFE, PARACHUTIST'S, SNAP BLADE.

OAL: 4¼ inches
Liners: steel
Bolsters: steel

This is an update of the knife shown on the previous page. The handle shape has been simplified and the bolsters have been removed thereby reducing production costs. It bears great similarity to the knives on the following pages, which are issue knives. It is not known to me that these single blade black handled knives were ever military issue but they were clearly the origin of others that were.

SCHRADE
WALDEN
NY USA

OAL: 4¼ inches
Liners: steel
Bolsters: n/a

all knives

OAL: 4¼ inches
Liners: steel
Volsters: n/a

In 1958 the MC-1 Survival Knife was adopted. It had a hook blade added to it to enhance its ability as a shroud line cutter. This knife has been made by Schrade, Camillus, and Logan/Smyth.

CAMILLUS
NEW YORK
U S A

SCHRADE
WALDEN
N.Y. U.S.A.
STAINLESS

LOGAN/SMYTH
VENICE, FL.
U.S.A.

CAMILLUS
NEW YORK
U S A

OAL: 3¾ inches
Liners: brass
Bolsters: steel

This is one of those knives that is hard to categorize as World War II or post-war. It bears a mark that was in use at the end of the war and may still be in use today.

This electrician's knife was made by Utica Cutlery Company of Utica, New York. The wood handle is stamped "TL-29" and filled with gold paint. The knife is marked as shown at the right.

UTICA
CUTLERY CO.
UTICA, N.Y.
MADE IN U.S.A.

OAL: 3¾ inches
Liners: brass
Bolsters: steel

It would be nice to find all your collectible knives in this condition ... new in the box. Unfortunately, it seldom happens. In this case we have a box of brand-new condition wood-handled TL-29s.

CAMILLUS
NEW YORK
USA

In the 1950s, the use of plastics was expanding and improvements were continuous. Early plastic knife handles often show signs of shrinkage and deterioration. Those problems had evidently been resolved by the time this knife was made.

After the war plastic replaced wood as the handle material of choice. For the first several years of its use it was marked with the "TL-29" in the same fashion as the wooden handled knives. Later, the marking of "TL-29" on the handles was dropped as is the case with the knife on the bottom.

all knives

OAL: 3¾ inches
Liners: brass
Bolsters: nickel silver

CAMILLUS
NEW YORK
USA

CAMCO
USA

For a few years Camillus Cutlery Company produced these TL-29s. They were made of thinner material and metal stampings. They were, altogether, a cheaper knife. Gone was the "TL-29" stamping on the handle along with the etched closing instructions on the screwdriver blade. The main blade is etched, "PROPERTY OF U.S. GOVT."

OAL: 3¾ inches
Liners: brass
Bolsters: nickel silver

This Camillus electrician's knife with black plastic handles bears the numbers that associate this knife with the Navy's Bureau of Aeronautics. The knife shown at the bottom of the page was made by Boker, probably in the 1980s.

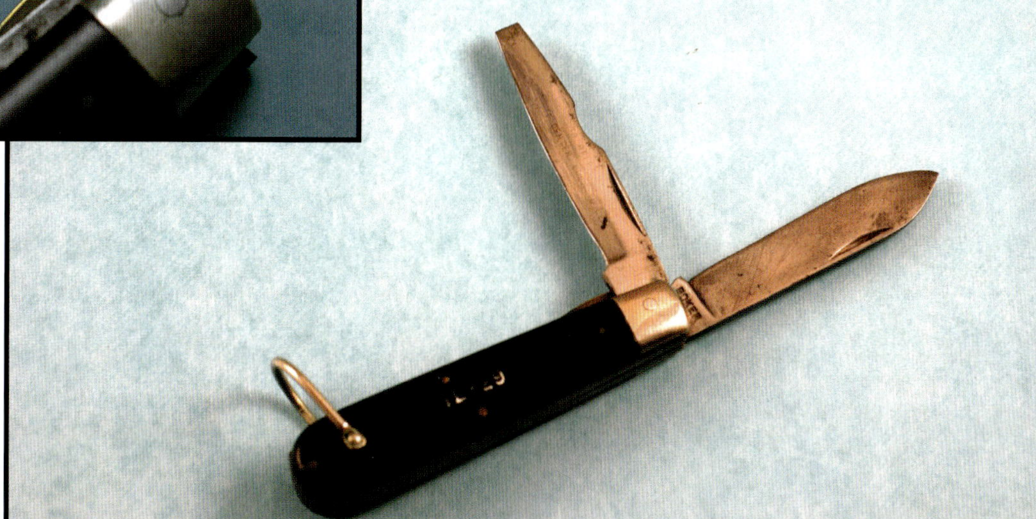

BOKER
USA

This Imperial TL-29 was most likely made in the 1960s when the blades were marked with the government ownership designation. The handles are black to red mottled plastic. Below are three TL-29s by Ulster. The bottom knife is marked with the handles stamped TL-29 and as shown in the inset. The center knife is only marked on the blades as shown in the inset while the top knife has only the makers mark on the tang and "U.S. Govt./TL-29 Specification" etched in the blade center.

This is an unusual TL-29. Its handle is stamped with the usual "TL-29" but the maker's mark is as shown at the right. The closing instructions on the screwdriver blade are stamped into the blade rather than etched. This is the first foreign made TL-29 I have encountered. The knife is very similar to U.S.-made knives. It is 3¾ inches long and has brass liners and steel bolsters. It is unknown whether this was a military contract item.

"ATCO"
ITALY

After World War II we had many troops stationed in Europe and it appears that we may have let contracts to local suppliers. The TL-29 above is marked, "Rich A. Herder/Solingen/Germany." Etched on the screwdriver is, "To release push center lock to left." The liners are steel and so is the piviot pin which is typical of German construction. Bolsters are of nickel silver.

Another foreign made TL-29 is shown above. Its handles are wood and the side liners and bolsters are steel. The center liner is brass. The blade tang is marked "J.H. Dotzert/Frankfurt A/M." The screwdriver blade is deeply etched "TO RELEASE PUSH CENTER LOCK TO LEFT."

The TL-29 stamping on the wooden handles is in quotes. The blade tang is marked ROMO with an arrow through it. The reverse of the tang is marked Italy. The screwdriver blade is stamped "TO RELEASE PUSH CENTER LOCK TO LEFT." There is some conjecture that this knife and others like it were made for the commercial market and capitalized on the popularity of the military TL-29 and may have never been military contract items.

Here is another Italian TL-29. Like the one before it, it has the designation in quotes. Also like the previous knife, the lock release information is stamped into the blade. The blade tang is marked "Unimart/Italy."

The "KNIFE, POCKET, GENERAL PURPOSE," can provide the collector with an almost endless opportunity for collecting and study. The knives above are, from left to right, made by Ulster, Camillus/1958, Imperial/1966, Camillus/1973, and Camillus/1976. Besides the World War II versions other maker's names like Queen and Schrade have graced these knives. Many of these knives have been sold commercially and some have been made by firms not supplying the military. Those that display "U.S.M.C.," "U.S.N.," or "U.S.A.F." on the handle have no real military affiliation.

OAL: 3¾ inches
Liners: steel
Bolsters: n/a

CAMILLUS
NEW YORK
U S A

KUTMASTER
UTICA, N.Y.
MADE IN U.S.A.

OAL: 3⅝ inches
Liners: brass
Bolsters: nickel silver or stainless

Just when you think that the all metal "KNIFE, POCKET, GENERAL PURPOSE" had taken over, a couple more Engineer's knives turn up. These are products of the 1950s and are made by Camillus (top) and Kutmaster (bottom). Similar knives by Ulster have been observed. Main blades are etched "PROPERTY OF U.S. GOVT."

CAMILLUS
NEW YORK
U S A

CAMILLUS
NEW YORK
U S A

This engineer's knife by Camillus was most likely made shortly after World War II while the Air Force was still part of the Army and known as the Army Air Force (AAF). This knife is found with both the Camillus four line mark and their more recent three line mark as shown above.

KUTMASTER
UTICA, N.Y.
U.S.A.

OAL: about 4 inches
Liners: brass
Bolsters: nickel silver or stainless

These three blade stockman's pattern pocket knives were made by Kutmaster. The lower knife with the black handle is the earlier knife. Both were probably made in the 1950s or 1960s.

KUTMASTER
UTICA, N.Y.
U.S.A.

This Kutmaster three blade utility knife proudly displays the United States Navy initials down the center of the main blade. This differs from the previous Kutmaster utilities shown in that the handles are genuine jigged bone.

OAL: about 4 inches
Liners: brass
Bolsters: nickel silver or stainless

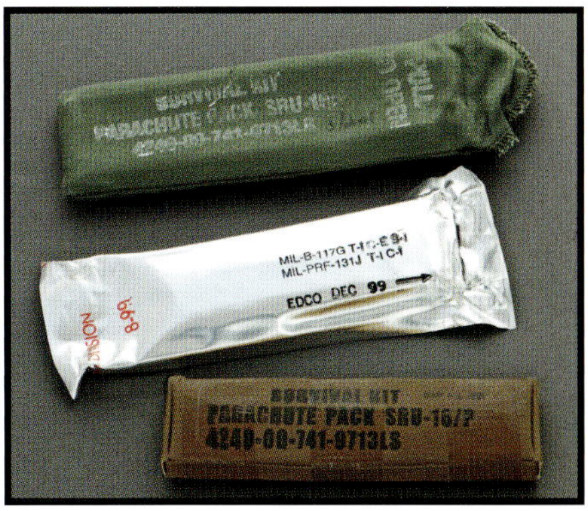

SRU-16/P

Some parachute packs since the 1980s contained these small survival kits called the "Parachute Pack SRU-16/P." The kit measures about 5½ inches long by 1½ inches wide by ¾ inch deep. The items are enclosed in a cardboard box sealed within a foil pack in a cloth cover. The items are shown and identifed below. Of interest is the small three blade pocket knife. This one is marked "Jaguar" and made in China but I have seen earlier packs with similar knives made by Colonial Cutlery Company.

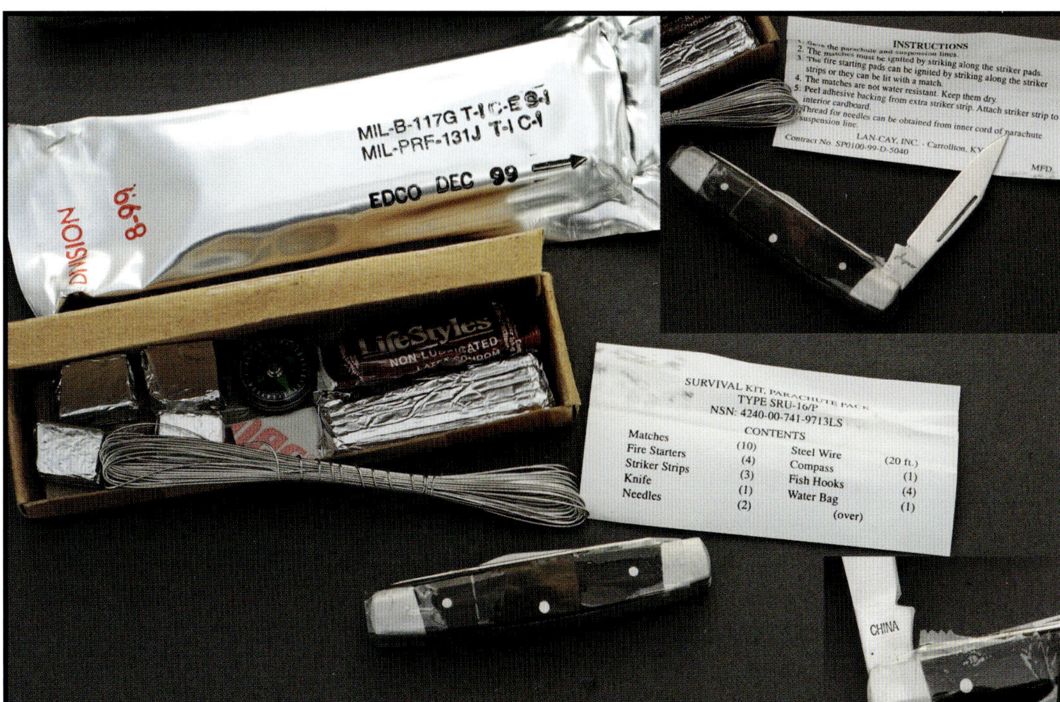

SORSAKOSKI FINLAND

Although technically not the United States military, the United States Central Intelligence Agency (CIA) did issue these knives in Vietnam in the early-1960s to its operatives and Vietnamese friendlies. This is a design based on the butterfly knife executed by Tapio Wirkkala in Sorsakoski, Finland. The knife is 4¾ inches long closed and has plastic handles. Later models can be found with red or black handles that are marked "Hackman Finland."

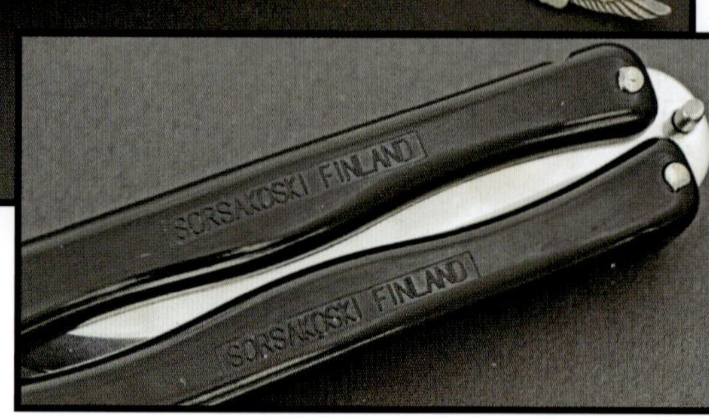

CASE

This Case Cutlery Company trapper model was issued by the CIA in 1971 to a U.S. Army Special Forces sergeant in Laos who was then on Project 404. This project was a covert operation supplying line crew technicians in support and training of the Royal Laotian Air Force while at the same time providing Raven Forward Air Controllers who supplied piloting expertise and guidance for running tactical air force operations. The Case knife is a standard 5¼ inches long closed. Blades have been blackened.

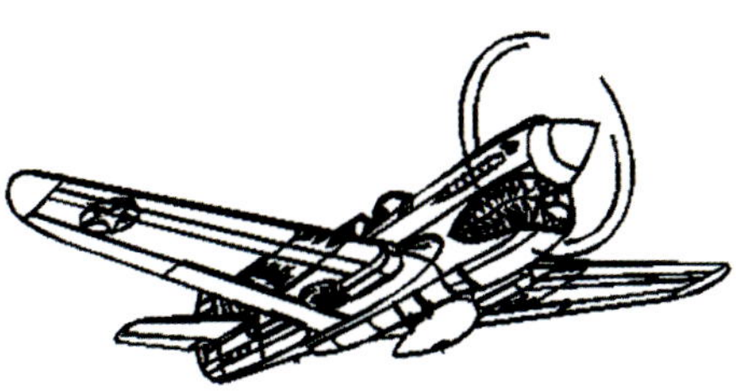

Souvenir knives were available in the larger cities of Vietnam and here are a pair of nearly identical knives. They differ only in color, one being black and one green. The lockback stainless steel blades are about 4½ inches long. Handles are made of decorated plastic. Undoubtedly, even some of these rather delicate knives saw rough service and few have survived to remind their owners of their time in-country.

This small unmarked knife (about 2¾ inches long) was originally World War II Japnaese issue. A supply of these was found on Okinawa in the early-1960s as we were outfitting South Vietnamese as well as our own personnel for cross border operations. Such "black" operations were originally conducted by the CIA but in 1964 MACV-SOG took over their control. The small knives with a can opener were useful in the field and had the added feature of being "unattributable" because they were unmarked.

Mad Magazine

C4 KNIFE

This is most likely the rarest of the post-war pocket knives. It is an experimental non-sparking knife made for working with explosives. Camillus Cutlery Company made about 200 of these knives, which were issued in 1967. The knife is 5½ inches long closed and has three blades. Both the main blade and the short screwdriver blade lock in the open position by way of a liner lock.

The C4 knife is 5½ inches long when closed and has a blade of 4½ inches. The main blade and the spike are made of non-sparking materials, a beryllium copper alloy. The screwdriver and backspring are steel. Frank Trzaska managed to get some government information on this knife declassified and wrote an article about the knife in the January, 1998 edition of *Knife World*. Among other things, he found out that just about 180 of these knives were ever produced, thus making this one of the rarest military knives ever.

THE BLACK KNIFE

The knife shown below is a prototype workbench sample of the Black Knife. It was one of the early attempts to work out the problems with the geometry of the blade and lockup arrangement as well as the overall design. You can see that the finished product had some refinements before it was ready for issue. This robust knife was well accepted by those who had reason to carry it.

This large automatic knife measures 5½ inches closed and found popularity with special operations forces beginning in the late-1980s. The knife saw service with the Navy SEALs in Operation Desert Storm. The knife is completely unmarked and was made by Ochs Enterprises in Florida. There are no liners and the handle is made of anodized aluminum. This is as high tech as it gets for now.

The knives shown below are all three blade stockman's pattern pocket knives and were all issued within the past several years. From left to right the makers are Kutmaster, Camillus, Ontario Knife, and Queen Cutlery. All are about 4 inches long and have brass liners and nickel silver or stainless steel bolsters.

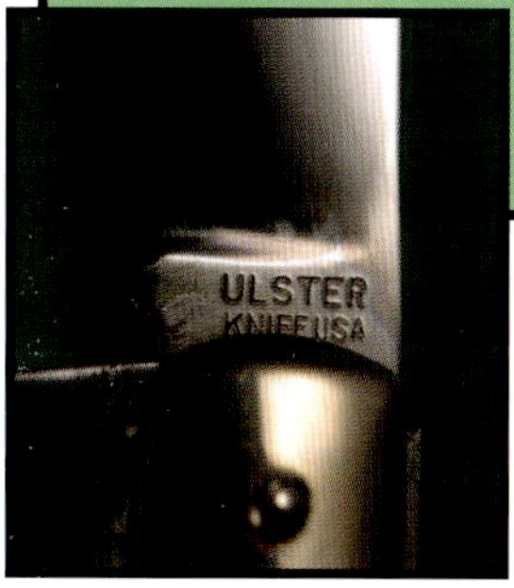

When Schrade Cutlery Company was closing its doors for good, this knife was purchased at the factory auction. It was made as an experimental pattern in the 1940s and later in the 1990s adopted as the standard pattern for the military rope knife. It is built on the same handle pattern as the World War II rope knife made by Imperial Knife Company. The knife by Camillus shown on the following page was the contract rope knife for the military through the 1990s and early-2000s.

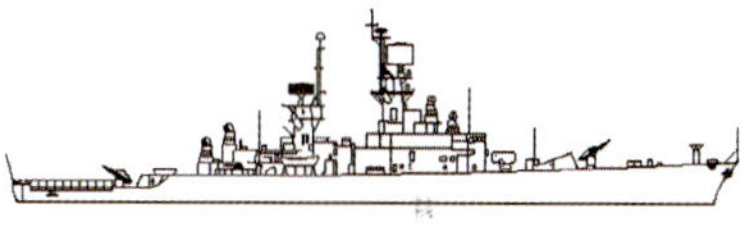

CAMILLUS
STAINLESS
U S A

Another fairly recent addition to the inventory of military knives is this rope knife by Camillus that was made for the United States Coast Guard. This particular knife was made in the 1990s. It is 4⅜ inches long and has liners of brass and the bolsters are nickel silver or stainless steel.

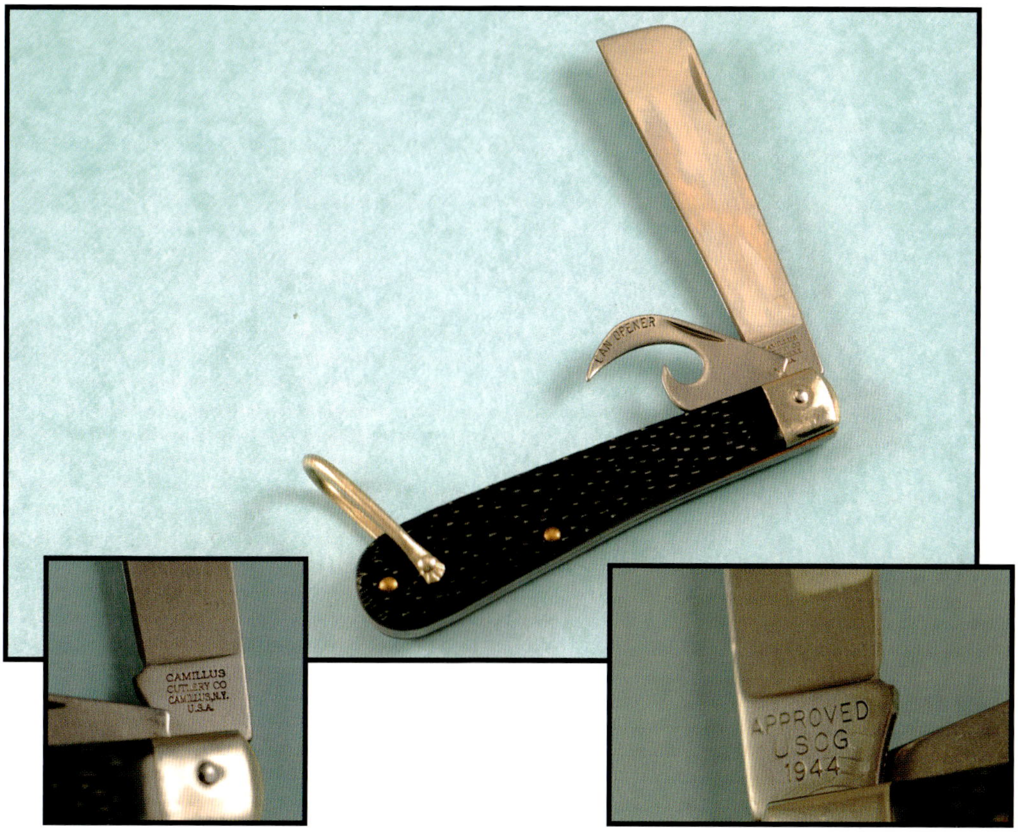

This is a modern knife, made in the 1990s, of the type shown on the previous page. These have been issued to sailors throughout the service. This knife, however, is of interest because it has a World War II blade despite the remainder in contemporary fitting.

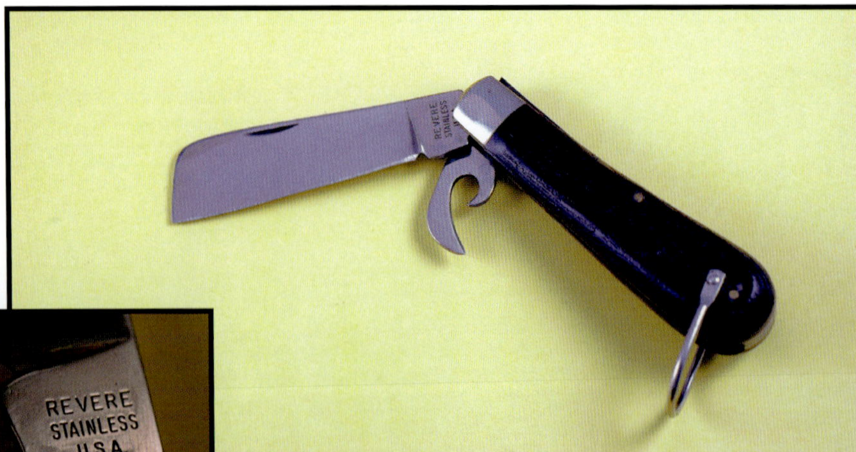

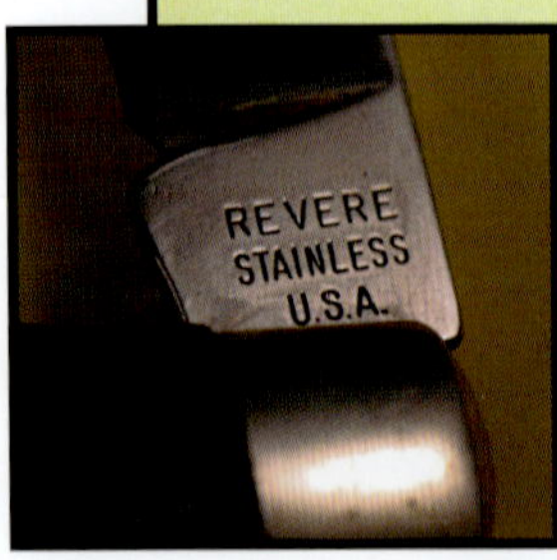

These are the latest generation of Navy rope knives. The knife shown above is made by Revere and like all the latest knives it has a can opener as well as the rope blade. The knife shown below is made by Datrex and is the same configuration as the previous knife. The tang is stamped DX1510M. Both these knives have substantial handles ... thick and heavy. The Revere knife has brown wood-grained plastic while the Datrex knife has smooth black plastic.

This is another Datrex-made military rope knife but the quality is not as good as the knife shown on the previous page, which was made in the U.S. This specimen was made in China. The date on the packaging of this knife is May 1996.

COLONIAL
PROV. R.I.

This is the 21st century version of the MC-1 Survival Knife commonly issued to pilots and paratroopers. This version utilizes a coil rather than a flat spring to activate the main blade. The shroud line cutter hook blade is manually opened. The knife is manufactured by Colonial Knife Company.

OAL: 4½ inches
Liners: none
Bolsters: none

Red Cross Knife

These small tool knives were given to U.S. soldiers serving in Iraq by the Red Cross. The knife measures 2¾ inches long when closed not including the key chain attachment. Tubular tool to the right is a flashlight. The knife is completely unmarked.

The CIA Multitool

One of the earliest multitool knives in the military was the OSS Escape Knife of WWII. It should be no surprise then that a modern version would find a place in that honored organization. The new knife has many more uses than did the old one and it seems that early on in Afghanistan a CIA Bravo Team Medic carried one of the Schrade ST-1 tools and used its double cut saw to perform amputations on wounded Afghan soldiers. This same medic was the first to reach Mike Spann after he was killed November 25, 2001. (Spann was the first American killed in combat during the U.S. invasion of Afghanistan in 2001. He was a paramilitary operations officer in the former Directorate of Operations, Special Activities Division (SAD) and was killed during a riot at the Qala-i-Jangi compound in Mazar-i-Sharif, northern Afghanistan.) This is the Schrade Tool ST-1, made in the U.S. Schrade USA has been out of business since 2003, and the tool is now manufactured in China and marketed by Taylor Cutlery as the Schrade Tool Model 'ST1N' and it is marked "China" inside the handle.

Leatherman Multitool

Oregon attorney Bob Lowry heads an organization called the Oregon Posse, a group whose members actively support our troops in Iraq and Afghanistan by supplying them with what they need and can't get through regular military channels. Items supplied by the Posse might include Swisher Sweets Cigars, Jif peanut butter, Blistex, paint balls and wrist rockets, beef jerky, ballistic sun glasses, chewing tobacco, candy, soccer balls for the kids in the streets, high candlepower lights, DVDs, magazines, and any number of the things we take for granted here in the States. The Posse deals directly with unit commanders to ensure that the shipments get into the hands of the men. Most communication is via the Internet and packages are shipped via USPS Priority Mail. From my observations, the members of this group have been extraordinarily generous with their support and the troops have been singularly grateful. Always responding to specific troop requests, it is noted that almost nothing tops the list as frequently as do multitool knives, particularly favored by the helicopter pilots who are comforted by having one in the survival pocket of their flight suit. For more than a year now Bob Lowry and the Oregon Posse have been purchasing the Leatherman Wave with cap crimper and shipping these to the men in Iraq and Afghanistan. This is the blackened model 830489 with nylon sheath. Although not an official government issue item plenty of these have seen service in mountains and deserts of the Mid-East.

Standard Issue Multitool

I obtained the first of these issue pieces from a returning Air Force enlisted man who was given the multitool in Iraq. It is the Gerber model MP600 needle nose. It is complete with the nylon belt sheath and many features including: pliers, wire cutter, wire crimper, straight edge blade, serrated edge blade, Phillips screwdriver, small, medium, and large screwdrivers, lanyard ring, can and bottle opener, file, and ruler. This is a later issue, which is the same model, but it has been upgraded to include tungsten carbide inserts for the wire cutter feature and the straight edged blade has changed from a clip point to a Wharncliffe pattern. Today in 2009, these multitool knives seem to be everywhere and in an endless variety of configurations. They have become the indispensable "tool box" that a soldier can carry on his belt or web gear. There is little doubt that more of these will find their way into military circles in days to come.

APPENDIX:

FALSE ASSUMPTIONS

The top knife is marked Kutmaster and the lower knife is marked Wards. On the shield is the mark of the Medical Department of the United States Navy. At one time I considered these to be genuine issue knives but I now believe them to be fantasy pieces made in the 1990s and early-2000s. There are a number of other patterns with similar markings and bearing the MD-USN shield. All are in new condition. I believe that they were put together from old parts. Many were sold on eBay.

LOGOS

MILLER
BROS
U.S.A.

IMPERIAL
PROV, R.I.

MILLER BROS

ULSTER
KNIFE Co.
U.S.A.

EMPIRE
WINSTED
CT.

MADE IN
USA

CAMILLUS
CUTLERY CO
NEW-YORK

PARKER
& FIELD

TL-29

UNITED
G.R. MICH.

ENGLAND

CAMILLUS
NEW YORK
USA

PAL BLADE CO.
MADE
IN U.S.A.

A WILLIAMS
SHEFFIELD. ENG.

"ATCO"
ITALY

MILLER
BROS

<> R <>
RUSSELL & CO
GREEN RIVER
WORKS

ULSTER
KNIFE Co.
U.S.A.

H.H.TAYLOR
&BROTHER
SHEFFIELD

MILLER BROS
CUT. CO
MERIDEN

JUNIOR
ARMY & NAVY

MADE IN
U.S.A.

KINGSTON

CAMILLUS
NEW YORK
USA

SCHRADE
WALDEN
N.Y. U.S.A.
STAINLESS

APP'VD 1944
U.S.C.G.
#85C

U.S.A. Remington MADE IN

COLONIAL
PROV. R.I.

UTICA
CUTLERY CO.
UTICA, N.Y.
MADE IN U.S.A.

E.C.SIMMONS
KEEN
KUTTER
ST LOUIS
MO

CASE
TESTED XX

WARDS
MADE IN U.S.A.

APPROVED
U S C G
1944 Q5

CAMCO
USA

CATTARAUGUS
CUTLERY Co.
LITTLE VALLEY
N.Y.

CAMILLUS
CUTLERY CO.
CAMILLUS.N.Y.
U.S.A.

ULSTER
TL29
U.S.A.

U.S.A. REMINGTON MADE IN

MADE IN
U.S.A.

UTICA
CUTLERY CO.
UTICA, N.Y.

ROBESON
U.S.A.
CUTLERY

ALFRED
WILLIAMS
SHEFFIELD
ENGLAND

E.J.HIGGINS
CASTSTEEL

CHALLENGE
CUTLERY.Co
SHEFFIELD

CHAPIN & PHIPPS
CELEBRATED
CUTLERY

AW

WARDS
MADE IN U.S.A.

H.H. TAYLOR
& BROTHER
SHEFFIELD

THOMAS
TURNER&Co
SHEFFIELD

WARNE
CHEEVER & CO

KUTMASTER
UTICA, N.Y.
U.S.A.

TALLY- HO
C.T. BINGHAM
SHEFFIELD

THOMAS FENTON

SCHRADE
CUT.CO.
WALDEN,N.Y.

CAMILLUS
NEW YORK
U S A

SCHRADE
WALDEN
NY USA

CAMBRIDGE
CUTLERY WORKS
SHEFFIELD

ALFRED
FIELD & Co
SHEFFIELD

ULSTER
KNIFE Co
NY

KUTMASTER
UTICA, N.Y.
MADE IN U.S.A.

GEORGE
WOSTENHOLMS
I·XL CUTLERY

J. WARD & Co
BRONXVILLE
NEW YORK

CAMILLUS
CUTLERY CO.
CAMILLUS, N.Y.
U.S.A.

ULSTER
KNIFE USA

CAMBRIDGE
& Co
SHEFFIELD

LOGAN/SMYTH
VENICE, FL.
U.S.A.

ALFRED WILLIAMS
SHEFFIELD, ENGLAND

SCHATT & MORGAN
CUTLERY CO.
TITUSVILLE PA

MILLER BROS
CELEBRATED
CUTLERY

F.WESTPFAL
ACME
ENGLAND

MILLER BROS
CUT. CO
MERIDEN

APPROVED
U.S.C.G.
1944-796
KUTMASTER
UTICA,N.Y.,U.S.A.

H.B.CLAFLIN
& COMPANY

DATES TO REMEMBER

1. Integral liners and bolsters not before 1860
2. Liner lock from 1900
3. Nickel silver from 1800
4. Bone stag, pickbone from 1800
5. "Prussia" before 1915
6. "Occupied Japan" 1945 to 1952
7. Stainless steel after 1916
8. Country of origin mark after 1890
9. Aluminum handles not before 1886
10. Cast steel 1700 to 1925
11. Shear steel to 1900
12. Celluloid not before 1870
13. Etched tangs not before 1880
14. Frosted etch on blades not before 1940
15. Hard rubber (gutta percha) start in 1850
16. Bakelite 1925 to 1950
17. Spanish American War 1898
18. WWI 1914 to 1918
19. WWII 1939 to 1945
20. Civil War 1861 to 1865
21. Vietnam War 1962 to 1975

BIBLIOGRAPHY

Bannerman Catalogue of Military Goods - 1927, (reprints), DBI Books, Inc., One Northfield Plaza, Northfield, Il 60093, 1980.

Bazelon, Bruce S. and McGuinn, William F., *A Directory of American Military Good Dealers & Makers 1785 - 1915, Combined Edition,* published by REF Typesetting & Publishing, Inc., Manassas, Virginia, 1999.

Brunner, Dr. John W., *OSS Weapons,* published by Phillips Publications, P.O. Box 168, Williamstown, N.J. 08094-0168, U.S.A., 1994.

Cole, M.H., *U.S. Military Knives Bayonets & Machetes Book III,* published by M.H. Cole, 501 Ridge Road, Birmingham, AL 35206, third printing 1983.

Cole, M.H., *U.S. Military Knives Bayonets & Machetes Book IV,* published by M.H. Cole, 501 Ridge Road, Birmingham, AL 35206, 1990.

Flook, Ron, *British and Commonwealth Military Knives,* published by Airlife Publishing Ltd, 101 Longden Road, Shrewsbury, SY3 9EB, England, 1999.

Goins, John & Charlotte, *Goins' Encyclopedia of Cutlery Markings,* published by Goins'Antique Knives, P.O. Box 33525, Indianapolis, IN 46203, 1998.

Hughes, Gordon & Jenkins, Barry, *APrimer of Military Knives European & American Combat, Trench & Utility Knives Part One,* published by Gordon Hughes, 121 Preston Drove, Brighton, England, 1973.

Hughes, Gordon & Jenkins, Barry, *A Primer of Military Knives European & American Combat, Trench & Utility Knives Part Two,* published by Gordon Hughes, 121 Preston Drove, Brighton, England, 1973.

Levine, Bernard, *Levine's Guide to Knives and Their Values,* fourth edition published by DBI Books a division of Krause Publications, Inc., 700 E. State St., Iola, WI 54990,1997.

Military Knives a Reference Book, published by Knife World Publications, P.O. Box 3395, Knoxville, TN 37927.

Peterson, Harold L., *American Knives the First History and Collectors' Guide,* published by Charles Scribner's Sons, New York.

Platts, Harvey, *The Knife Makers Who Went West,* published by Longs Peak Press, Inc., 1651 South Sunset, P.O. Box 1762, Longmont, Colorado 80501, 1978.

Stephens, Frederick J., *Fighting Knives An Illustrated Guide to Fighting Knives and Military Survival Weapons of the World,* published by Arco Publishing, Inc., 215 Park Avenue South, New York, N.Y. 10003, 1985.

Tweedale, Geoffrey, *The Sheffield Knife Book A history and Collectors's Guide,* published by The Hallamshire Press, Interleaf Productions Limited Broom Hall, Sheffield S10 2Dr, UK, 1996.

INDEX

N

O

P

Q

R

S

T

U

V

W